W9-BVX-326

BRACING

FOR

ARMAGEDDON?

BRACING

FOR

ARMAGEDDON?

THE SCIENCE AND POLITICS
OF BIOTERRORISM IN AMERICA

WILLIAM R. CLARK

OXFORD
UNIVERSITY PRESS
2008

OXFORD
UNIVERSITY PRESS

Oxford University Press, Inc., publishes works that further
Oxford University's objective of excellence
in research, scholarship, and education.

Oxford New York
Auckland Cape Town Dar es Salaam Hong Kong Karachi
Kuala Lumpur Madrid Melbourne Mexico City Nairobi
New Delhi Shanghai Taipei Toronto

With offices in
Argentina Austria Brazil Chile Czech Republic France Greece
Guatemala Hungary Italy Japan Poland Portugal Singapore
South Korea Switzerland Thailand Turkey Ukraine Vietnam

Copyright © 2008 by Oxford University Press, Inc.

Published by Oxford University Press, Inc.
198 Madison Avenue, New York, NY 10016

www.oup.com

Oxford is a registered trademark of Oxford University Press

Library of Congress Cataloging-in-Publication Data
Clark, William R., 1938–
Bracing for armageddon? : the science and politics of bioterrorism
in America / William R. Clark.
p. cm.
Includes index.
ISBN 978-0-19-533621-4
1. Bioterrorism—United States. 2. Bioterrorism—United States—Prevention.
3. Biological weapons. I. Title.
HV6433.35.C53 2008
363.325'30973—dc22   2007040445

1 3 5 7 9 8 6 4 2

Printed in the United States of America
on acid-free paper

Acknowledgments

I would like to extend my profound thanks to a number of individuals who shared their time and expertise with me during the preparation of this book: Dr. Milton Leitenberg, Center for International and Security Studies, School of Public Policy, University of Maryland; Dr. Jeanne Guillemin, Boston College and Senior Fellow, MIT Security Studies Program; Dr. Alan Pearson, Biological and Chemical Weapons Program, Center for Arms Control and Non-Proliferation; Dr. Laurene Mascola, Chief, Bioterrorism Preparedness Unit, Los Angeles County; Mr. Frank Simione, Vice President of Management and Compliance Services, American Type Culture Collection; Dr. Ray Zilinskas, Center for Nonproliferation Studies, Monterey Institute for International Studies; Dr. Seth Carus, Deputy Director, Center for the Study of Weapons of Mass Destruction, National Defense University; Dr. James G. Hodge, Jr., Executive Director, Center for Law and the Public's Health, Johns Hopkins University; Dr. Amy Smithson, Center for Nonproliferation Studies, Monterey Institute of International Studies (Washington, D.C.); Dr. Bryan McDonald, Center for Unconventional Security Affairs, University of California, Irvine; Dr. Susan Wright, History of Science and International Relations, University of Michigan.

Preface

Over the past two decades, an enormous effort has been mounted by numerous federal and state agencies to prepare America to defend against the possibility of a bioterrorist attack. This effort jumped ahead at warp speed following the horrendous World Trade Center and Pentagon attacks of September 11, 2001, followed by the postal anthrax scares just a few weeks later. Five people died in these latter incidents, considered by some to be the opening salvos in a new form of terrorism brought to our shores. By the end of 2008, the United States will have spent nearly fifty billion dollars upgrading almost every conceivable aspect of our ability to respond defensively to a catastrophic bioterrorism attack.

Concerns about bioterrorism in America, while certainly justified in many respects, have at times and in some quarters risen almost to the level of hysteria. Part of the reason for this is doubtless the conflation of bioterrorism with a larger "war on terror." Declaring war on something is a time-honored way in American politics to raise an issue to a level of unquestionable urgency. Another part of the terror of bioterrorism is that, unlike terrorism using other weapons—bombs, chemicals, nuclear devices— bioterrorism is based on things we cannot see and few of us understand. We rely on scientific experts to explain them to us, adding yet another layer of uncertainty, both for the public and for our political leaders. Science is not always objective, and scientific experts themselves have differing points of view—political points of view—about bioterrorism, just as they have differing points of view about global warming, or stem cell research, or the beginnings of life.

America in fact faces bioterrorism from not just one but two sources: from other humans who would use deadly biological weapons to attack us for political or ideological reasons, and from nature, which periodically unleashes biological pathogens on us for reasons that have nothing to do with human affairs, but that terrorize us nonetheless. We remain prey to infectious disease pandemics like the 1918 influenza outbreak, which killed perhaps a hundred million people worldwide, and the yet unresolved threats posed by the viruses causing SARS and H5N1 avian flu.

Both of these biological threats to our security and well-being, the political and the natural, are real. History makes clear that natural bioterrorism—pandemics—will, like earthquakes, tsunamis, and hurricanes, absolutely recur periodically, with the potential for catastrophic physical, social, and economic damage. What has been lacking in our approach to the threat of political bioterrorism to date is an assessment of exactly how real it is. How likely is it that America will experience a bioterrorist attack that could bring this country to its knees? What would it take to mount such an attack? Who could do it, and what weapons would they use? How would bioterrorism compare with the damage America would suffer from other forms of terrorism, or from a natural pandemic like that of 1918? How much of our resources as a nation should we spend preparing for each? How do we know when we are safe, or at least as safe as we can realistically expect to be, from these threats? These are hard questions.

Given the numerous instances of outright terrorism over the last two decades of the twentieth century and on through September 11, 2001, it may not be surprising that our perception of the threat of bioterrorism has at times verged on the hysterical. Fear is not necessarily a bad thing; hysteria is almost always counterproductive. Fear is a response dictated by nature to ensure survival, and as we will see in the chapters that follow, we have responded vigorously to nature's command. Not always rationally, but certainly vigorously.

But it is time now to move toward a more mature view of bio-terrorism, to tone down the rhetoric and see it for what it actually is: one of many difficult and potentially dangerous situations America—and the world—face in the decades ahead. What have we actually gotten for the tens of billions of dollars we have spent so far on bioterrorism defense? Was this money well spent? No nation has infinite resources, and we must accept that we will never be able to make ourselves completely safe from every threat we face. We will have to make rational assessments of those threats we can identify, and apportion our resources as intelligently and effectively as we can to deal with them. In this book we look at the scientific, political, legal, and social facets of bioterrorism that can guide us as we attempt to bring this particular threat into a more realistic perspective for the twenty-first century.

CONTENTS

BRACING

FOR

ARMAGEDDON?

CHAPTER I

TALES OF A DARK WINTER

A Play in Three Acts

THE FOLLOWING "PLAY" WAS PERFORMED IN THE SUMMER of 2001. It may seem somewhat dated from our perspective in 2008, especially with the departure from the world political scene of Saddam Hussein. But it had a major impact on the thinking—visceral as well as intellectual—of those charged with considering how America should respond to the possible threat of bioterrorism. It is worth watching this play once again. Admission is free.

ACT I. DECEMBER 9, 1999. 7:00 PM

Members of the National Security Council file into a brightly lit conference room for what promises to be a long and difficult meeting. It will certainly be that, but not for the reasons they think.

Most members attending this evening have spent the better part of the day organizing their thoughts about the items on this evening's agenda, several of which are going to require a clear U.S. response in the days ahead. Tension has been rapidly escalating between China and Taiwan. China recently test-fired medium-range ballistic missiles by arching them over Taiwanese airspace. The Taiwanese are reacting furiously. But since then there has been another peculiar and potentially dangerous incident. A number of pig farms in Taiwan have experienced serious outbreaks of foot-and-mouth disease. Rumors are circulating in the Taiwanese press and other media that this was a deliberate biological weapons attack mounted by China against Taiwanese agriculture. China is vigorously denying the accusation, but tempers are rising on both sides.

In what looks like a potential coup, the FBI has worked with Russian police and intelligence agencies to arrange a "sting" operation that netted a senior Al-Qaeda operative as he was attempting to purchase fifty kilograms of plutonium in Russia. This individual had also made inquiries about obtaining certain biological warfare weapons produced some years earlier by Soviet Union laboratories. The United States needs to craft a careful plan about how much of this to make public and how much to keep under wraps. The Russians are already beginning to leak information that many on the U.S. side want kept classified.

And Saddam is at it again. After the lifting of military and economic sanctions against Iraq six months ago and the ending of the no-fly-zone restrictions, Hussein has been going all out to beef up his military, across the board. There is now hard intelligence that he has imported materials that could be used for chemical and biological weapons. Moreover, at least three former Soviet biological weapons specialists are known to be in Iraq, and presumed to be working at Iraqi weapons facilities. The Joint Chiefs of Staff are very worried about this development. Most worrying of all, however, Iraqi troops and equipment are starting to filter off toward the border with Kuwait. That country, together with Bahrain and the United Arab Emirates, is requesting deployment of Western forces in the region to forestall another 1991 adventure on the part of Saddam Hussein. A coherent strategy, involving several key allies, will have to be formulated quickly.

There may not be time to get to the other issue pressing on everyone's mind and time: Y2K.

As the President's national security advisor rises from his seat next to the President to signal the start of the meeting, the room falls silent. He has a videocassette on top of the stack of papers in front of him. He and the President had been in hushed, heads-together conversation since they came into the room, and the advisor looks decidedly more solemn than usual. We can sense people in the room sitting forward ever so slightly in their chairs as he begins to speak.

"Before we begin this evening, I'd like you to see a video-tape we just copied in the last half hour of news coming out of Oklahoma City. This is from a local station, but I expect it will be on all the major networks, including CNN internationally, before the hour is out."

He hands the cassette to an assistant, who moves to a tape machine near a large TV screen at one end of the room. The security advisor resumes his seat next to the President.

The images that come on the screen are mostly of two Oklahoma City news anchors. Someone keeps switching between them and shots taken outside a large brick hospital. It isn't clear if these latter are supposed to be live shots or file footage. A picture of a man in a white coat comes on briefly, then flashes off again. He looks like a doctor, but says nothing—the station is obviously trying to assemble a story on the fly.

Back to the anchors. In the past few hours, it appears one person has been diagnosed—"with a high degree of certainty"—with smallpox. That in itself would be a major story, since no one has been diagnosed anywhere in the world with smallpox for over twenty-five years. But there are now rumors circulating that as many as a dozen more persons are under surveillance in area hospitals, suspected of having the same disease. The Centers for Disease Control and Prevention (CDC) in Atlanta have been alerted, and we learn that officials there are scrambling to get flights to Oklahoma City.

A low buzz goes around the room as members of the council begin whispering to each other. The tape plays out, offering little hard information about what kind of response is being mounted locally. No doctors or any other health authorities speak during the remainder of the tape.

The security advisor signals for the TV to be turned off. He sits down, and the President stands up and looks around the table.

"In view of the potential extreme urgency of this situation, I am setting tonight's agenda temporarily aside. I talked just a few minutes ago with the Secretary of Health and Human Services, who confirms that we do indeed have smallpox in Oklahoma City. In fact the latest word I have is that there may be cases showing up in Pennsylvania and in Georgia as well, although these cases have not yet been confirmed to me by the Secretary.

"This is extremely serious. As I'm sure most of you know, small-pox no longer exists in nature. As I understand it, it's just about impossible that this case—these cases, it now appears—arose either naturally or by accident. There is no smallpox virus stored anywhere near Oklahoma City. We are at this point assuming the virus was introduced deliberately, by foreign state- or non-state-sponsored operatives. We are considering this a bioterrorist attack on the United States."

There are audible inhalations, and more than a few paling faces, around the table. The President pauses for a moment to let this sink in, then pulls some handwritten notes from his jacket pocket and spreads them in front of him.

"I'm told that much of the United States is extremely suscep-tible to smallpox. Vaccinations were halted some thirty years ago. Those who were vaccinated before that time probably have only minimal resistance to the disease today. The same is true of much of the rest of the world. We anticipate that this news, once circulated, will cause considerable alarm, probably panic, in many places. Mostly here at first, but likely spreading abroad as well.

"I'm also told that given the incubation period for this virus, the attacks responsible for the cases we're seeing now probably took place somewhere between one and two weeks ago. No coun-try or group has yet claimed credit for this attack. The FBI and CIA have no leads at the moment. What's going on may be related to decisions we have made or are considering making with respect to the current situation in the Middle East. It could be intended to distract the leadership of this country, or to intimidate our civilian population. We just don't know at this point.

"Governor Keating of Oklahoma has been in Washington this week on state business, and agreed to come here tonight to brief us on what he knows about the events in his state. He's under-standably eager to get back to Oklahoma City. Governor?"

The Governor states that Oklahoma City hospitals are already experiencing overflow from persons worried about their health status, and this seems likely to mushroom in the days to come. At the same time, some hospital personnel are leaving their posts to be with their own families. This could very quickly lead to a health care crisis. Emergency disaster plans are being implemented as quickly as possible. The Governor says he has implored the CDC to release smallpox vaccine; he would like to be able to assure his state's 3.5 million residents they will all receive immunizations against smallpox. He ends by stating he has declared a full-blown state emergency, and requests the President to immediately declare a federal state of emergency in Oklahoma as well.

At this point an assistant to the Secretary of Health and Human Services (HHS) stands up to present a hurriedly pre-pared background talk on smallpox. He confirms that the last natural case of smallpox in the United States occurred in Texas, in 1949. Public health officials ceased offering smallpox vaccina-tions to the general public in 1972. Best estimates are that at least 80 percent of the current U.S. population has no effective protec-tion against smallpox. The United States at present has about 12 million reliable vaccine doses. Some portion of that has been set aside for the U.S. military, which might be called upon to go into areas where chemical and biological weapons could be used.

Smallpox is highly contagious, he says, passed from person to person by aerosols—small, germ-laden droplets breathed or coughed into the air. The average number of new persons infected by an uncontrolled infected person is estimated at ten, but could be as high as twenty. Isolation of infected patients and immunization of their contacts can help greatly in controlling spread of the disease. The disease itself, once it sets in, is untreat-able. The historical mortality rate from smallpox in this country is about 30 percent. Current hospital bed capacity in the United States could probably not handle more than a 10 percent increase in patient load.

A member of the Council raises his hand. "Twelve million doses of vaccine. That seems like a lot, but probably isn't. Do we know how much of that is committed to the military?"

The Secretary of Defense responds from across the table. "We have requested that 2.5 million doses be held in reserve for pos-sible use by troops deployed to the Middle East. We have every reason to believe that at least Iraq, and possibly other countries, may be in possession of weaponized smallpox virus. God knows what else."

Another question from the table for the HHS representative. "And how do you propose to use the rest?"

"That's something we hope you will discuss here this evening. Basically there are three general strategies for distributing drugs or vaccines in major medical crises, depending on supply. In the 'ring' strategy, enough vaccine would be provided to immunize emergency personnel in the affected area, plus known contacts of the primary victims. The second option is to ship enough vaccine to the affected cities to treat their entire populations, holding the rest in reserve until we see what happens. The third option is essentially the second option, except that the remainder of the vaccine would be distributed immediately, on a per capita basis, to states not yet affected. They could use it as they see fit."

"And a follow-up question here. Do we have any idea of how many people were exposed?"

"Actually, we don't. And because it happened probably about ten days ago, we may never know. We'll just have to wait and see how many cases develop."

The Attorney General raises his hand. "I can see some tricky legal questions down the road here. Who's going to be running the show as we develop a response? What is the federal government's role in this situation, aside from providing vaccine? What are the states' roles? Who's going to coordinate all this? Who has final authority, and where?"

FEMA Director Hauer responds. "My office certainly has a lead role in coordination of efforts at all levels. But I already see a number of points that will need clarification...."

An aide slips into the room to hand the President a memo, which indicates that now there are twenty confirmed smallpox cases in Oklahoma City. Fourteen other cases in that city are suspected, and first reports from Philadelphia and Atlanta show seven and nine cases suspected in those two cities. The CDC now has people on-site in Atlanta and Philadelphia, and about to arrive in Oklahoma City. The CDC has 100,000 vaccine doses prepared for immediate shipment to each of the three cities.

As the First Act ends, the Council members begin discussing what the government should do and collecting questions that will need answering as soon as possible. The lights go down, and we see in our program notes that they debated for another hour and a half, closing with a strong appeal to the President to make a public announcement as soon as possible and to follow up with regular reports to the public as long as the emergency lasts. The people will need to be willing allies of the government, and the government will need to earn their confidence with full and frequent disclosures of exactly what is happening to them.

ACT II. DECEMBER 15, 1 PM

The Deputy National Security Advisor begins the meeting by reviewing television and press coverage of the smallpox crisis in recent days. Many of the fears expressed at the first meeting are apparently being more than realized.

Over 2,000 cases of smallpox have now been reported in fifteen states. At least 300 people are already dead, and mounting numbers are reaching a terminal stage of their disease. Congress is demanding swift action to seek out and punish those responsible.

Hospitals in several cities are completely overwhelmed, partly from the increased caseloads of patients who need to be isolated, and partly from the fact that a large number of workers are refusing to enter hospitals for fear of catching the disease and passing it on to their families and friends.

Isolated cases of smallpox have shown up in several foreign countries; all are traceable to one of the three primary outbreak locations in the United States. Some countries are already closing their borders to U.S. goods and to U.S. citizens unless they have proof of recent vaccination.

Ghastly pictures of smallpox victims, including children, are routinely appearing in all media outlets, further driving the mounting panic evident everywhere. These images are being broadcast worldwide, including on Al Jazeera.

Vaccine distribution has been chaotic, and vaccine doses that have been shipped from CDC are being used up at an alarming rate. Riots have broken out near some distribution centers, requiring police and National Guard troops to guarantee the safety of public health workers.

In Oklahoma, the governor has closed all schools and colleges and cancelled all sporting events for the foreseeable future, as well as other public gatherings. He is considering closing shopping centers and certain large businesses as well. The economic toll stemming from such action is unfathomable, but public safety may well require it. Food shortages are beginning to show up in some of the harder-hit areas, as truck drivers refuse to enter them.

The President takes over the meeting to present a few additional brief remarks on the current situation.

"As far as any of us can determine, the events we are now experiencing, which are spreading rapidly, all stem from the three primary attacks in Oklahoma, Pennsylvania, and Georgia, somewhere around December 1. Neither the FBI nor any of our intelligence agencies has yet developed any hard leads on the origin of these attacks, or who might have carried them out. We are completely convinced this was not a case of domestic terrorism. The only supplies of the smallpox virus in the United States are in the CDC in Atlanta, and I have been assured and reassured that these are under extremely strict control. The facility housing them has been examined, and there is nothing at all to suggest these stocks have recently been disturbed. So I think we can put domestic bioterrorism to rest as an explanation for what happened.

"Our intelligence people feel the most likely source is some of the smallpox virus originally stored in Moscow, and maintained there after the collapse of the Soviet Union. Our people have long worried about the security of that stockpile. We have no reason to believe the Russians themselves had any hand in this, at least as a state. But we are concerned that some of their scientists who have access to these materials may have taken some of the virus with them when they moved to other states looking for employment. Many rogue states would dearly love to get their hands on biological agents such as smallpox, and we believe that former Soviet scientists work in a number of these states.

"Dr. O'Connor from Health and Human Services will now update us on the status of our response to these attacks. Dr. O'Connor?"

"Thank you, Mr. President. A detailed epidemiological analysis carried out by our staff indicates that all three initial attacks took place in crowded shopping malls, probably on a weekend. Exactly how the virus was released is unknown at present.

"The number of new cases per day is leveling off at present, but as second- and third-round infections begin to show up—any day now—we expect these to rise, and very sharply.

"We are rapidly running out of vaccine. A million doses were sent to each of the three primary attack states. Half a million doses are being sent to each of twelve other affected states. Two million doses have been set aside for the military. That leaves us with just over a million doses. We expect that to be gone within a day or two. Gearing up for new vaccine production, even waiving the normal FDA procedures in this case, will take months. So we have no hope of producing our own vaccine in the very near future. We have contacted other countries for loans from their existing stocks. Great Britain has pledged half a million doses; Russia is considering a loan of five million doses.

"We have issued guidelines on the use of existing vaccine stocks, asking that they be used only to immunize first responders—police, fire, National Guard, hospital personnel, HazMat units—as well as verified immediate contacts of confirmed victims. But ultimately that is a decision for individual states. There is a shortage in all states of individuals willing to help trace down patient contacts for potential immunization, for the same reasons workers refuse to go into hospitals. We understand pressures on distribution systems and personnel have been enormous, and we have no idea whether our guidelines are being adhered to."

The Attorney General interrupts. "Can we get clarification on some of these issues? We're providing the vaccine; certainly we have the authority to say how it is to be used?"

"I suppose if we were to set up federal vaccination centers in each city, we could control how the vaccines are administered," the President replied. "But short of that, I don't see what we could do. But let's get CDC and HHS to chime in on this." He nodded to the Secretary of Health and Human Services, and signaled for Dr. O'Connor to continue.

"Health care systems in some of the harder-hit cities are being rapidly overwhelmed. Refusal of health care workers to enter hospitals remains a severe problem. As far as I know these people have all been offered the vaccine, but the response has not been encouraging.

"There are not nearly enough isolation facilities to house infected patients. The National Guard is helping to erect military hospital tent units to house smallpox patients, but at some point we may run out of those. Also, we are finding that many families are trying to move relatives with non-smallpox problems out of hospitals treating smallpox patients.

"We are seeing misinformation being passed along—presumably unintentionally—by media of all types. For example there have been several charges that people in poorer sections of cities are not getting equal access to vaccinations. But we are also seeing outrageous Internet scams trying to sell drugs for treating smallpox. Such drugs simply do not exist. HHS has established a secure website with accurate information about smallpox that we hope the public will use.

"Schools are now closed almost nationwide. This is putting an additional strain on families with two working parents. Public gatherings of all kinds are also being prohibited. These steps are essential for controlling spread of the disease, but some are already beginning to question their necessity in light of the extraordinary disruptions they cause. People are also being asked to voluntarily restrict their travel. This may be working, but it is possible that in the near future such restrictions will have to be made mandatory."

"What about quarantining?" the Attorney General asked.

"It's being carried out irregularly. There is clear authority for quarantining in essentially every county of every state, but people are proving to be extremely resistant. It's one thing to confine the occasional person with whooping cough to their house for a few days, and nail a sign on their door. It's another, psychologically, to put up barriers around entire sections of a city."

"I suppose we should have prepared people for this in the first day or two of the outbreak."

"Yes, I agree. That's about all I have to say for now. Thank you all for your attention. I'll pass you along now to Mr. Hauer, Director of FEMA."

"Thank you, Dr. O'Connor. I have only a few brief remarks. FEMA is also receiving reports that the health care systems in a number of states are reaching a crisis point. Delivery of health care to non-smallpox patients is seriously eroding. In Oklahoma, for example, twenty of 138 hospitals have closed their doors to further admissions of any kind. This is probably illegal, but that is a state matter.

"The National Guard in many states is doing an outstanding job of providing security and delivering supplies where commercial truckers won't go, but their numbers are not sufficient to do everything that's needed. The Red Cross has set up numerous shelters with food and water, but like hospitals they're having a hard time finding people to staff them.

"The National Disaster Medical System is not working particularly well in this situation. The system calls for hospitals in unaffected areas to accept overflow patients from crisis regions, but the last thing we want to do is transport active smallpox patients long distances and implant them in unaffected areas. The thirty-person volunteer medical teams that are part of the system have provided only modest support over the past week. Again, many volunteers are proving reluctant to go into heavily infected areas, where they are needed most. Also, many of these people have been co-opted by their home states.

"We are now seeing individual states beginning to close their borders with surrounding states. This raises some real issues with respect to the national interstate highway system and interstate commerce, which is a federal issue. We need to discuss what we can do to keep these highways open, and not just to federal traffic, which probably won't be interfered with. This is already having a disastrous effect on the shipping of food and goods, with economic consequences we can all imagine. I hope the Council will address this issue quickly, or refer it immediately to appropriate agencies.

"I believe that is all I have to say for the moment."

ACT III. DECEMBER 22

The meeting opens with another review of recent events as reported in the media, presented by the Deputy National Security Advisor.

"We now have over 15,000 active cases of smallpox in twenty-five states. The vast majority of these have come up in just the past few days. The CDC anticipated this, and made sure media do not portray this as additional attacks. We are seeing the spread of the virus from the initial round of infections earlier this month. There are now smallpox outbreaks in at least ten other countries traceable to the outbreaks here.

"All media are reporting the mounting economic damage from the initial attacks. There are reports that international investors and banks may be liquidating their positions in American financial instruments, real estate, and businesses.

"As casualties mount from the second wave of infections, particularly in Georgia and Pennsylvania, we see many people moving out of those areas, trying to find places free of the disease. Some of those people are bound to be infected, and this may hasten the spread of disease to new areas. In Atlanta and Philadelphia most businesses and banks are now closed.

"Public opinion polls show about half of Americans would support an all-out nuclear attack on any foreign country found to be responsible for these attacks. The same polls show tremendous anger with government at all levels for failure to halt the epidemic.

"Canada and Mexico have now completely sealed their borders with the United States. Nearly all countries worldwide are refusing entry to American citizens, even with proof of recent smallpox vaccination. American goods are being turned back from all foreign ports of entry.

"We are starting to see armed clashes between police and National Guard units and civilians, and among civilians themselves. Some people have violently resisted removal of clearly infected family members to isolation units. Others have attacked mortuaries and morgues, trying to recover remains of loved ones to avoid mandatory cremation. In a number of cases individuals have used armed force to drive away persons they suspect of being infected.

"The President will now say a few words."

"Thank you, Randy. I will announce today that I have already signed an executive order waiving normal FDA testing procedures so that U.S. pharmaceutical companies can immediately begin production of smallpox vaccine. Every step will be taken to assure safety of these vaccines, but I have granted the companies involved immunity from civil or criminal prosecution should the vaccines cause harm. We must do this, or risk catastrophic consequences. The companies, the FDA, and the CDC will closely monitor those to whom the new vaccines are given for signs of any problems. The bad news is that although the companies will all work twenty-four-hour shifts if necessary, we will not see the new vaccine for at least five to seven weeks.

"I just want to remind you that as we focus much of our energy and attention on the smallpox epidemic, and how best to use scarce federal resources to help states with their battles on that front, we must also discuss urgent questions arising from the worsening situation in the Middle East. I may have more to say about that later in this meeting. We are all somewhat reassured to see that the China-Taiwan situation seems to be cooling off, at least for now.

"I will now ask Dr. O'Connor to give us an update from HHS and CDC on progress of the epidemic. Dr. O'Connor."

"Thank you, Mr. President. A few days ago, we had about 2,000 total smallpox cases. Today we have 16,000 cases, from which we can expect around 5,000 fatalities. We anticipate another 30,000 cases over the next week. In another two-three weeks, as we see third-round cases start to appear, we could see as many as 300,000 active infections, with 100,000 deaths. Our smallpox vaccine supply, including loans from Great Britain and Russia, was exhausted two days ago. As the President just indicated, we will be slogging through this crisis without vaccine for at least five weeks—well beyond the anticipated third round.

"The magnitude of the burst of second-round cases represents a failure to identify, immunize, or isolate infected persons before they passed the virus along to others. The third-round burst we expect in the coming weeks will represent a failure to identify second-round victim contacts and to vaccinate or isolate them before they passed it on. This is rapidly becoming the number-one issue in containing this epidemic, and as the number of infections expands we simply may not be able to keep up with it. We estimate the 16,000 cases confirmed as of this date will have made on average a hundred contacts during their contagious stage, before they were diagnosed and isolated. That's 1.6 million contacts to trace down and deal with. We probably have manpower to deal with less than a tenth of that."

The Director of FEMA looks up. "Where do you see this ending?"

"That depends on many factors, mostly on how quickly we see new vaccine production and how many doses will be available over what time span. Our current projections, which assume new vaccine production stays on schedule and that the vaccine is fully effective, are that by mid-February we will have seen—conservatively—a total of three million cases of smallpox, with an anticipated total of one million deaths."

An aide to the President enters the room and hands him two messages. The room falls quiet as the President reads these. He rises to speak.

"This morning, three American news outlets received identical anonymous letters demanding that all U.S. troops be pulled out of the Middle East, and all U.S. warships leave the Persian Gulf. If we do not do this, there will be additional attacks, this time with anthrax and plague in addition to smallpox."

The President looks around the silent table.

"That's not all. Each of the letters also contained a printout of the genetic blueprint of a virus. This was immediately faxed to the CDC, which as you know has sequenced the blueprint of the virus recovered from our smallpox victims."

He holds aloft the second message.

"The two blueprints are identical. This is very, very real."

The final curtain comes down, and the house lights go on.

So where was this play performed? Did anyone ever see it?

This "play" was performed only once, across the days of June 22–23, 2001, a little over eleven weeks before September 11. Only the cast and about fifty or so government officials ever saw it. It wasn't a play, of course. It was a government exercise called Dark Winter, one of a number of such exercises carried out to test America's ability to respond to major health emergencies arising from terrorist-caused disasters involving radiological, chemical, or biological weapons (Box 1.1). These exercises have probed interactions and communications among responding agencies

BOX 1.1

EXERCISES IN BIOTERRORISM

Exercise	When	What
TOPOFF I[1]	May 2000	Plague
Dark Winter	June 2001	Smallpox
Sooner Spring	April 2002	Smallpox
TOPOFF II	May 2003	Plague
Atlantic Storm[2]	January 2005	Smallpox
TOPOFF III	April 2005	Plague

This is a partial list of such exercises. Several of these also tested responses to chemical and radiological attacks.

[1]TOPOFF is an acronym for "Top Officials."
[2]Atlantic Storm was an extension of Dark Winter to an international stage. Brad Smith et al., "Navigating the Storm: Report and Recommendations from the Atlantic Storm Exercise," *Biosecurity and Bioterrorism* 3(2005):256–267.

and their senior leaders, adequacy of communication across lines of authority, the ability of public health officials to respond at the local level and of hospitals to mobilize emergency supplies and generate beds, and many other aspects of how government agencies at various levels would function during a response to a catastrophic event.

Dark Winter was mounted by a formidable array of producers: the Johns Hopkins University Center for Civilian Biodefense Strategies, the Center for Strategic and International Studies, the Analytical Services Institute of the Department of Homeland Security, and the Oklahoma National Memorial Institute for the Prevention of Terrorism.

Only some of the "actors"—those presenting information about the terrorist attack and its sequelae—actually had scripted lines. The other players—about a dozen in all—were real-life, senior-level policy makers, drawn from various parts of the federal government to play the roles of key government officials and members of the National Security Council. (The Governor of Oklahoma, Frank Keating, was actually recruited to play himself.)

It was enacted at Andrews Air Force Base, outside Washington, D.C. Most of the participants were simply asked to react to facts pouring in to them in the form of mock news releases and videos, and to factual information, assessments, and estimates provided orally or as written memoranda.[1] They were asked to make immediate decisions that would shape the response of the federal government to the events unfolding first in Oklahoma, and shortly across the entire nation. After extensive discussion, and calls for additional information or clarification, they sketched out a series of recommendations for presentation to the President. Decisions and recommendations made at the end of one "Act" (NSA meeting) were incorporated on the spot by the producers into the scripting of subsequent meetings. The NSA members also had to participate in a mock press conference with five real-life journalists partway through the exercise.

The responses of the government officials taking part in this exercise were exhaustively analyzed by the "producers," and a number of conclusions were drawn: Senior-level decision makers were unfamiliar with bioterrorism, and did not know what options were available to them in the event of an attack; information flow in an event of this type was totally inadequate; the United States had essentially no smallpox vaccine; the U.S. health care system was woefully unprepared to deal with a catastrophic health emergency, whatever its source; the respective roles, responsibilities, and authorities of the federal versus state governments in situations of the type were completely unclear.

No one involved in this exercise could have imagined that less than three months later America would experience a real-life terrorist attack of unprecedented magnitude. Partly, perhaps, as a reaction to exercises like Dark Winter, the government rushed special teams trained to detect the kinds of germs used in biological weapons to New York, Washington, and Pennsylvania to determine if the airplane attacks had been accompanied by release of deadly viruses or bacteria. None were found. But barely a month later came news that must have sent chills down the spine of everyone who had played even a minor role in Dark Winter. A person or persons unknown had disseminated deadly anthrax spores through the U.S. postal system. This attack would eventually kill five people and seriously injure a dozen more and cause substantial social and economic disruption.

Dark Winter and related exercises were intended to put a real scare into government policy makers and members of Congress. Together with the events surrounding September 11, there is no doubt they did. These exercises have been rightly criticized for presenting extreme scenarios, often inflating the capabilities of terrorists or the human threat of various pathogens.[2] Unrealistic statements about the threat posed by bioterrorism continue to this day, at the highest levels of government.

One result of this heightened state of fear, at best only marginally based in reality, is that an enormous effort has been mounted by numerous state and federal agencies to prepare America to defend itself in the event of a catastrophic bioterrorist attack. This effort has coincidentally contributed to a modest improvement of our ability to respond to nature's own form of bioterrorism: infectious disease pandemics like the 1918 influenza outbreak, which killed perhaps a hundred million people worldwide, and the yet unresolved threats posed by the viruses causing SARS and H5N1 avian flu. Almost everyone now agrees that the health threat posed by outbreaks of emerging pathogens is certainly as great as, and probably greater than, that posed by bioterrorism. Bioterrorism is often characterized as an event of low probability but high consequence. Pandemic influenza, like a magnitude 8 earthquake in California, is an event of near certainty, with unimaginable consequences (chapter 4).

At some level bioterrorism remains a threat, and no one is prepared to dismiss it out of hand. The question policy makers now face is: how much preparation is enough? How do we know when we are safe, or at least as safe from bioterrorism as we can reasonably expect to be? This is not an easy question. Since September 11, tens of billions of dollars have been spent upgrading almost every conceivable aspect of our ability to respond to a bioterrorism attack. The underlying programs continue to be sold to the Congress and the American public largely by playing on fears of bioterrorism.

But our resources are not infinite, and hard decisions have to be made about the relative risks posed by bioterrorism in comparison to other challenges America will face in the years ahead. In this book we will look at the scientific, social, legal, and political facets of bioterrorism that are shaping, and will continue to shape, decisions about how we deal with a much wider range of potentially terrifying threats facing us at the beginning of the second millennium.

CHAPTER 2

A BRIEF HISTORY OF BIOTERRORISM

EXERCISES LIKE DARK WINTER PROVIDED POLITICAL leaders with frightening insights into what a real bioterrorist attack might look like, but even more importantly, how woefully inadequate our ability was to respond to any large-scale health emergency, including a bioterrorist attack. As mentioned earlier, the exercises themselves have been rightly criticized as exaggerating certain aspects of what such an attack might look like, particularly in terms of the technological capabilities imputed to the terrorists.[1] But they certainly served to jumpstart a major upgrading of public health and emergency management services at all political levels, all across America.

Terrorism is a word rarely heard in English-speaking countries until a few decades ago. Modern political usage of the term stems from the French Revolution, one phase of which the French refer to as *la Terreur*. Current usage was undoubtedly shaped by the string of airliner hijackings in the Middle East in the 1960s and '70s, of which one partisan allegedly said "[The hijackings] did more for our cause than twenty years of pleading [before the U.N.]."

These hijackings were occasionally referred to by the media as terrorism, and the hijackers as terrorists. The rise of groups in the 1960s and '70s such as the Red Brigades, the Black Panthers, and the Baader-Meinhof Gang left no doubt that the use of terror and violence against civilians to achieve political ends was real. But certainly after the first bombing of the World Trade Center in 1993 and the 1995 destruction of the Murrah Federal Building in Oklahoma City, these words were used with increasing frequency in the United States. Since September 11, 2001, they have become a part of the daily American vocabulary.

Attention to a cause is usually a major objective of most terrorism. As one expert put it, for the most part "terrorists want a lot of people *watching*, not a lot of people *dead*."[2] Terrorism achieves its effects not simply through violence but through an apparent randomness, or at least unpredictability, of violence that can produce uncertainty, fear, even panic in targeted populations. There are many definitions of terrorism (Box 2.1), just as there are many means of carrying out terrorist acts. Terrorists are generally accepted to be groups operating independently of formally recognized states, although they may accept support from states. The phrase "transnational groups" is in current vogue for many terrorist organizations.

The major forms of terrorism of concern to contemporary governments, aside from organized assassinations and bombings, are those involving nuclear, chemical, and biological weapons, also referred to as weapons of mass destruction (WMD). The use of biological organisms or their toxins to sow terror in a civilian population is called bioterrorism.

Bioterrorism is an offshoot of biological warfare, and like most progeny it differs somewhat from its parent. The main difference is that biological warfare is a highly organized aggressive activity carried out by one state against another, usually through a military arm, using biological agents to kill, disable, or disorganize people to achieve a largely military goal. However, in spite of the fact that most industrialized countries had intense biological weapon development programs in the early and middle twentieth century, there are very few documented instances of the use of bioweapons in modern warfare. We will discuss reasons for this in later chapters.

Bioterrorism, while using many of the same agents and tactics as biological warfare, is a more ad hoc activity carried out by individuals or political groups against other political groups or

---

BOX 2.1
SOME OFFICIAL DEFINITIONS OF TERRORISM

United Nations	Any act intended to cause death or serious bodily injury to a civilian, or to any other person not taking an active part in the hostilities in a situation of armed conflict, when the purpose of such act, by its nature or context, is to intimidate a population, or to compel a government or an international organization to do or to abstain from doing any act.
U.S. State Department	Premeditated, politically motivated violence perpetrated against noncombatant targets by subnational groups or clandestine agents, usually intended to influence an audience.
U.S. Defense Department	The calculated use, or threatened use, of force or violence against individuals or property to coerce or intimidate governments or societies, often to achieve political, religious, or ideological objectives.
U.S. Federal Bureau of Investigation	The unlawful use of force or violence against persons or property to intimidate or coerce a government, the civilian population, or any segment thereof, in furtherance of political or social objectives.

---

states, with a mixture of objectives (Box 2.2). As with other forms of terrorism, harming or killing individual human beings may be only part of the aim. Among other goals, panic and its attendant social and economic disruption are valued outcomes, as are bringing media attention to a particular political cause and recruiting more individuals to that cause. These goals, it should be noted, do not require killing or harming mass numbers of people. Several coordinated, perhaps sequential, small attacks could be every bit as effective as a single mass attack.

BOX 2.2

BIOTERRORISM DEFINED

Bioterrorism is the intentional use of any naturally occuring microorganism, virus, infectious substance, or biological product, or any bioengineered component of any such microorganism, virus, infectious substance, or biological product, to cause death, disease, or other biological malfunction in a human, an animal, a plant or another living organism in order to influence the conduct of government, or to intimidate or coerce a civilian population.

(Based on the Model State Emergency Health Powers Act; see chapter 8)

It is important to distinguish between bioterrorism, as we have defined it so far, and what we might call "biocrime." Biocrimes may involve many of the same agents as bioterrorism and biological warfare, but the aims are always much narrower—short-term gains for a restricted number of individuals. Once the immediate gain is achieved, the activity ceases. For example, after the postal anthrax attacks in 2001 (which we will discuss below), there were literally thousands of hoaxes, where harmless white powder was mailed to a wide variety of individuals, for a wide variety of reasons. In many cases it was simply a joke, but given the anxiety caused to the intended target and the substantial investment of police time committed to investigating these incidents, they certainly qualify as biocrimes.

In other cases, for example white powder sent to abortion clinics together with threatening notes, the intent was clearly intimidation in promotion of an ideological goal, which could be considered bioterrorism. These distinctions become important when we try to assess the overall magnitude and threat of bioterrorism against civilian populations. As discussed below, we still cannot say whether the postal anthrax attacks were bioterrorism or a biocrime, because we don't know who carried out these attacks, and thus we do not know the motives.

Various forms of biological warfare have a long if occasionally crude history, including dipping arrowheads and spear points into rotting cadavers or feces, or lobbing entire diseased corpses over town or castle walls. The perpetrators obviously had little understanding of what they were doing, so it may be less than accurate to call this biological warfare. But they must have thought it worked, or they wouldn't have done it.

And once the basis for infectious diseases was uncovered at the end of the nineteenth century, it didn't take long before biological warfare became a highly precise science. By World War I, and on through World War II, virtually every major military power—including the United States—and even a few minor ones had created scientific research units dedicated to the development of ever more lethal biological weapons. However, with the exception of Japan during its occupation of China and Manchuria, there was no extensive use of biological agents against either military or civilian targets during World War II, although the Allies seriously discussed carrying out several such attacks against Germany. In the United States the War Department (precursor of today's Department of Defense) established a special biological warfare facility at Fort Detrick, Maryland. This facility still operates as the U.S. Army Medical Research Institute of Infectious Diseases, although it no longer develops biological weapons, focusing instead on defense against such weapons, among other activities.

Biological weapons, more than any other weapon type, bring out the fear factor when used for terror purposes. They are poorly understood by the public, they are invisible, and they are totally indiscriminate in their action. Attempts to co-opt the weapons and methodology of biological warfare for terrorism were probably inevitable, and they were predicted by numerous scientists and politicians in many different countries in the decades after World War II. They became a reality in the last two decades of the twentieth century. We will focus here on several of these events, which collectively have shaped most of the debate about bioterrorism in the United States.

The first use of a biological agent as a terror weapon in the United States involved the Rajneesh cult in Oregon, in 1984. The second major attempt involved another cult, Aum Shinrikyo, best known for its use of the nerve gas sarin in a Tokyo subway in 1995, though the group had also worked extensively with biological agents. And finally, the dissemination of anthrax spores through postal service facilities in the United States represents the most sophisticated attempt yet to adopt the tools of biological warfare to terrorism, in the United States or anywhere else.

THE RAJNEESH CULT, OREGON, 1985

In 1981 the Rajneesh cult,[3] founded by a displaced Indian mystic named Bhagwan Shree Rajneesh, purchased a 60,000-plus-acre ranch in north central Oregon, not far from the city of The Dalles (population 11,000). The Rajneesh commune soon grew to several thousand souls, who enjoyed various degrees of success in their search for peace and enlightenment, in an atmosphere of easy drugs and sex. But the Bhagwan clearly flourished. He accumulated ninety Rolls Royces, five private jets, and a helicopter.

Not content with having built a thriving community on their own land, cult members gained electoral control of the nearby small (population 75) town of Antelope in Wasco County. They named their new town Rajneesh and quickly converted it to their own needs and ends, to the utter disgust of the mostly retired locals. Soon, perhaps growing weary of life in such a small town, Rajneeshees began vying for seats on Wasco County boards and commissions. Reaction at the county level was mixed, but mostly negative. A few saw potential downstream benefits from the influx of money and reasonably educated people, but most shared the views of their compatriots in Antelope.

In mid-September 1984, a dozen people who worked in or had recently eaten in several restaurants in The Dalles became ill from food poisoning. One of these restaurants, a Shakey's Pizza franchise, was co-owned by a member of the Wasco County land-use board, but this raised no particular flags at the time. The number of victims grew over the following week, and the biological culprit behind it was soon identified by public health officials: *Salmonella enterica typhimurium,* a bacterium commonly causing food poisoning.[4] Everyone was treated with appropriate antibiotics, no one died, and the incident seemed to have subsided.

But a week later it was back. This time ten restaurants were involved. Local health services, including medical laboratories, were overwhelmed. The only hospital in The Dalles quickly ran out of beds. The number of persons who became ill soon exceeded 700, considerably beyond what might be expected in a community of this size for a normal outbreak of salmonella poisoning. The city called the Centers for Disease Control and Prevention (the CDC) in Atlanta for help. By the time help arrived, local health officials determined that most or all of the affected people had eaten at salad bars, and restaurants were immediately advised to stop serving salad. They did.

There followed an exhaustive investigation of all suppliers of salad vegetables and dressings to local restaurants. Everything came up clean; even the local water. Preliminary reports from state and federal health investigators stated that the poisonings were most likely caused by accidental incursions of salmonella into the food supply of the restaurants involved. Even the CDC felt that food handlers were the most likely source for introduction of the bacteria into the salad bars. Some locals, in particular another member of the Wasco County land use committee, believed the Rajneesh cult was somehow involved, but lacking any hard evidence or direction from health authorities, investigations into this possibility eventually fizzled out.

The involvement of the Rajneeshees became clear only as the result, a year or so later, of internal squabbles within the cult. The Bhagwan himself implicated some of his lieutenants in the affair, and called for a government investigation, after which he beat a hasty retreat to India. Authorities found abundant evidence at the commune of not only *S. enterica typhimurium* but a fairly sophisticated medical research laboratory and evidence that the cult had considered employing other deadly pathogens, including HIV—the AIDS virus. They had purchased salmonella essentially over the counter, from a Seattle scientific supply house. Among their intended victims, in addition to various county officials, was U.S. Attorney Charles Turner, the top federal prosecutor in Oregon. He was to be spared infection with salmonella. Cult members planned to shoot him. They failed. But they also intended to use their cultured salmonella to poison the Dalles water supply in the days before an upcoming election. The grand plan, as it turned out, was to reduce the number of Wasco County citizens able to vote, thereby increasing the influence of Rajneeshees. For various reasons, this never came about.

Probably no more than a dozen Rajneeshee leaders were fully aware of or involved in the salmonella poisonings. Seven cult members were ultimately indicted in various murder or attempted murder conspiracies. The cases against the conspirators came to a close only in 2005, when the last of these returned from self-imposed exile in Germany and surrendered to authorities.

While the Rajneesh incident gained national and international attention among those who had been predicting bioterrorism in America, it could be argued that what happened in Oregon was not so much a form of bioterrorism as a simple criminal attempt to manipulate a specific civilian population and its various civil agencies through malicious intimidation—a biocrime. There was no discernible political aim beyond an attempt by a few members to influence a single election and to expand their power and influence within the cult, and perhaps the intimidation or possible elimination of some individuals. Still, many of the features of bioterrorism were there: preparation and crude weaponization of a human pathogen, delivery of the pathogen to intended victims, and serious social and psychological disruptions in the targeted population.

AUM SHINRIKYO, JAPAN 1985–1995

If you missed this first wake-up call, some would say, your attention may have been sharpened a decade later by events in Japan, involving a messianic cult called Aum Shinrikyo ("Supreme Truth"). Aum Shinrikyo is known mostly for its release of sarin nerve gas in the Tokyo subway system in March, 1995. This attack killed twelve people and injured perhaps a thousand more, some seriously.[5] In a preliminary run-up to this attack, cult members had carried out a previous gassing in the provincial town of Matsumoto that killed seven.

But Aum Shinrikyo did not limit itself to nerve gas as an agent of terror. A subsequent detailed investigation into their activities in the years preceding the subway attack revealed that cult scientists had also endeavored to develop lethal biological terror weapons as early as 1990, preceding their work with sarin. The cult was completely broken up and eliminated immediately after the subway attacks, and never did manage to produce an effective weapon based on biological agents. But not for lack of trying.

The difference between the Rajneeshees and Aum Shinrikyo is like the difference between a mom-and-pop corner grocery and a nation-wide supermarket chain. Aum was founded in 1984 by Chizuo Matsumoto, the son of a tatami mat weaver. Shortly after starting Aum, he changed his name to Shoko Asahara. He was blind in one eye at birth, with only limited vision in the other eye, and was considered legally blind. In spite of his handicap—possibly in part because of it—in school he was aggressive and bullying. As a young adult he was a determined entrepreneur with little regard for the law. By his early twenties he had amassed a significant personal bankroll through a series of acupuncture clinics and meditation centers that were largely clever scams. After a period of rather cynical exploration of various spiritual movements, he decided to found one of his own. We will never know his motives for choosing this path; his spiritual and philosophical utterances were, throughout Aum's existence, utterly erratic and disorganized.[6]

With his otherworldly looks, facilitated by his near-sightless gaze and long, flowing black hair, Asahara became a mesmerizing, even charismatic figure. And Aum Shinrikyo quickly grew to become one of the larger and more influential of the "new age" organizations in Japan. Based on absolute, unquestioning adherence to the whims and dictates of Asahara, the cult eventually had a membership in the tens of thousands. Among the many thousands who lived in Japan, about 1,400 chose to transfer all of their assets to Aum Shinrikyo and live in Aum compounds. All of the leaders of the cult were drawn from these members.

At its peak in the early 1990s, Aum controlled hundreds of millions of dollars, had extensive real estate holdings in Japan as well as offices abroad (including one in Manhattan), and operated as many as twenty businesses, many of which were nothing more than fronts to foster the cult's illegal activities.[7] Even at the end, when the police finally raided Aum headquarters, they found eight million dollars in cash and twenty-two pounds of gold!

After a visit by Asahara and his staff to Russia in 1992, where they staged numerous revival-style public meetings and purchased a radio station to spread the Aum gospel, another thirty thousand members were added to the cult's rolls—exceeding the number of adherents at home. Within Japan itself, Aum cult members were drawn largely from the middle and upper-middle classes. Laborers and service workers had too few assets to be of interest to Aum's leadership. Most members were young, and a large number were college-educated. When interviewed after the cult's collapse, many confessed to a sense of alienation from what they perceived as Japan's materialistic, work-oriented culture, which left them little time to explore themselves and other ways of looking at the world. Most also craved a sense of community and belonging which they found lacking in their daily lives.

Importantly, many of those drawn to Aum Shinrikyo had backgrounds in science and technology, having graduated from, or at least having attended before dropping out to join Aum, some of Japan's best scientific programs and graduate schools. A brief look at a few members of Aum's scientific staff will give a sense of the quality of people they were able to recruit:

Masumai Uchiya—trained in organic chemistry at Tsukuba University.

Toro Toyoda—left a graduate program in particle physics at Tokyo University to join Aum.

Ikuo Hayashi—cardiovascular surgeon trained at Keio University and at Mt. Sinai Hospital in New York

Seiichi Endo—doctorate in molecular biology; did research in viral genetic engineering at Kyoto University before leaving to join Aum.

Hideo Murai—advanced degree in astrophysics from Osaka University; did research for Japanese steel company.

Aum also recruited from among their membership numerous mid-level technicians and laboratory workers with various levels of scientific expertise.

The research facilities built by Aum Shinrikyo for developing biological weapons were located in their main compound near the village of Kamikuishiki at the base of Mt. Fuji, in a building on the east side of Tokyo, and at Mt. Aso in Kyushu province. No expense was spared in ordering the most up-to-date scientific equipment for working with viruses and with bacteria and bacterial toxins.

One of the first biological agents they tried to weaponize was botulinum toxin, one of the deadliest human poisons known: three billionths of an ounce may be fatal in humans. This nerve toxin is produced by a common soil bacterium, *Clostridium botulinum,* and cult scientists attempted to isolate the toxin from bacteria gathered nearby and grown in culture flasks in their own laboratory. Following protocols developed in the 1930s and '40s, they attempted to harvest and concentrate the toxin from bacterial growth medium and convert it to a form that could be sprayed from a high-pressure nozzle. This material was actually used in attacks against the Japanese Parliament, portions of the Tokyo airport, and even the headquarters of the American Seventh Fleet in Yokohama. None of these attacks produced any casualties. A second generation of botulinum toxin was used in an attempt to kill the royal family during the wedding of the crown prince, again without results.

The cult's scientists subsequently turned their attention to production of weapons-grade anthrax. As we will discuss in the next chapter, anthrax is particularly deadly in its spore form: dry, hard, dust-like particles that are easily carried great distances by even the lightest of winds and can be inhaled to produce the most lethal form of the disease.

The cult used large fans to disseminate what they believed was a highly purified preparation of anthrax spores from the top of a building adjoining their laboratory in Tokyo. Aside from a sickening stench and a few cases of mild illness in people and pets close to the building, there were no effects of the attack. No cases of anthrax were reported or detected. Had they prepared a fully potent batch of infectious anthrax spores, and disseminated it properly, it is likely that thousands—possibly tens of thousands—of people would have been killed. Obviously that was their intent. A second attempt to spread anthrax, this time around the Imperial Palace, was carried out a few months later, again with no effect. Subsequent investigations suggested they had cultured a relatively harmless anthrax strain normally used for animal vaccinations (and thus by definition unable to cause disease), and in addition had likely failed to convert it to the more deadly spore form of anthrax.

Members of the cult's scientific staff traveled far and wide in their attempts to procure lethal biological weapons. During Aum's many trips to Russia, particularly after the fall of the Soviet Union, attempts were made to ferret out sources for every conceivable sort of weapon, including even jet fighters and nuclear bombs! In 1992, cult members traveled to Zaire after an outbreak of hemorrhagic fever, trying to find a source for the Ebola virus. They also apparently attempted to work with a rickettsial agent causing the dreaded Q fever. Fortunately, none of these efforts bore fruit.

All in all, Aum Shinrikyu carried out around a dozen terrorist attacks with biological weapons, every single one of which failed to produce significant casualties. We will examine some of the reasons for these failures in chapter 9. But first, let's look at the third wake-up call, one that, while more limited in scope than that provided by Aum Shinrikyo, is in many ways more chilling: the dissemination of anthrax spores through the U.S. Postal Service shortly after the terrorist attacks of September 11, 2001.

AMERITHRAX 2001

The dissemination of anthrax spores using the U.S. Postal Service in September and October of 2001 is the most serious attempt to carry out an attack on U.S. soil using a bioweapon. The FBI code name for the still ongoing probe into these attacks is Amerithrax. The Amerithrax attacks also provide our first real glimpse into what might happen when determined terrorists produce, or otherwise acquire, high-grade human pathogens and use them effectively. These attacks killed five people and injured at least sixteen others, some of them seriously.

As we will discuss in more detail in chapter 3, anthrax spores can cause disease in two different ways, depending on how they enter the body. If the spores settle on exposed skin, and if the skin is even slightly damaged at the point of contact, the spores can penetrate the skin and cause *cutaneous anthrax*. This form of anthrax in humans responds well to currently available antibiotics, and fatalities are now almost unheard of. On the other hand, if spores are inhaled and enter the body through the lungs, *inhalation anthrax* will ensue. This form of anthrax is difficult to treat, and more deadly. None of the persons contracting cutaneous anthrax in the 2001 attacks died. Nearly half of those with confirmed inhalation anthrax died. Certainly many others came into contact with anthrax spores during Amerithrax, but for a variety of reasons (unabraded skin; timely administration of antibiotics) did not develop detectable infections.

As far as we know, the letters containing anthrax spores were mailed in two batches from New Jersey. The first batch, probably containing five letters, was postmarked September 18, 2001, and sent to news organizations in Florida (American Media, Inc.) and New York (ABC, CBS, NBC, and the *New York Post*). Only the letters to the *New York Post* and NBC were recovered for forensic analysis. A second batch of at least two letters was mailed on October 9. One was received in the office of Senator Tom Daschle (D-SD), and a second addressed to Senator Patrick Leahy (D-VT) was found a few weeks later among undelivered Senate mail. Both of these letters were recovered for scientific analysis.

No harm was done to the individuals to whom the letters were addressed. The targeted individuals rarely opened their own mail. With one or two possible exceptions, the victims of these attacks were office personnel or other associates of the primary targets, or people working in postal facilities through which the contaminated letters had been processed. Whether this "collateral damage" was intended or even envisioned by the terrorist or terrorists is unknown.

As a result of the attacks, one Senate office building was closed for several months for decontamination. Two major East Coast postal sorting facilities were closed for several years during clean-up procedures. This kind of social and economic disruption must be included when we add up the toll of attacks with bioweapons.

After more than five years of intense investigation by the FBI and postal inspectors, over 9,000 interviews, and the issuance of over 6,000 subpoenas, investigators still do not know (or at least are not saying) who carried out these attacks. For that reason, we do not know if they were truly bioterrorism or simply a biocrime. But because of their proximity in time to the World Trade Center and Pentagon attacks of September 11, initial focus was on foreign terrorists. It seemed a reasonable assumption that the two incidents could be related.[8] It now seems more likely that one or more highly skilled American scientists were responsible, directly or indirectly, for at least the production of the anthrax used, and probably for its dissemination. The FBI leaked the name of one scientist who has been the subject of an intense but inconclusive investigation for nearly six years. An undisclosed number of other "persons of interest" are said also to be under active investigation.

The anthrax spores used in the Amerithrax attacks distinguish these incidents from all others we have experienced. Unlike the harmless vaccine strain used by Aum Shinrikyo scientists, at least some of these spores were from a highly virulent form of anthrax, and milled into a fine powder ideal for dissemination by a number of means. FBI tests showed that the spores had been generated in the past two years but did not indicate where. Had Aum Shinrikyo had spores of this quality, events in Japan might have taken an entirely different turn.

The strain of anthrax used in these attacks is maintained mostly in university and defense institute research laboratories, and is very difficult (today impossible) to obtain casually. All experts now agree that the spores found in at least the Senate letters were very pure and highly potent. It is less certain whether they had been treated with chemical reagents that would maximize airborne dispersal. The experts are also nearly unanimous that the spores could only have been produced in a very high-tech, fully equipped laboratory by a person or persons expert in every aspect of growing anthrax bacilli and converting them to spores. The spores may have been stolen from one of these facilities and disseminated through the mails by an unskilled person, but they were very unlikely to have been made by a non-scientist in a garage laboratory, and certainly not in a cave in Afghanistan.

There are numerous genetic substrains of anthrax, and these can be identified by their individual DNA fingerprints. It is known where each of these substrains was produced and by whom. But prior to the Amerithrax incidents, researchers readily exchanged their strains among themselves, keeping at best casual records of their exchanges, and often none. Every laboratory in the United States that works with anthrax, including in particular the government's own Fort Detrick facility in Maryland, has been extensively investigated by the FBI. Many foreign laboratories have also cooperated in this investigation. But to date, the FBI has been unable to identify (or at least has not done so publicly) the exact source of the anthrax strain used in the 2001 attacks.

Amerithrax provided us our first—albeit modest—glimpse of what bioterrorism in America might look like. It is not entirely clear that this was in fact a bona fide bioterrorist attack, since we don't yet know the motives of the perpetrator(s). But it played another role whose importance cannot be overstated. A growing chorus had been slowly developing in the United States over the preceding decade, warning—with only partial success—of the dangers of bioterrorism. The juxtaposition of Amerithrax and exercises like Dark Winter against the background of this chorus and the attacks on the World Trade Center tilted the American political mind down a path we are still treading. We would spend tens of billions of dollars over the next eight years trying to find our way.

Between the Rajneesh episode of 1984 and the 2001 postal service anthrax attacks, there were two other incidents involving the use—or at least the acquisition for presumed intended use—of biological weapons by U.S. citizens. Although most ordinary Americans barely noticed nor long remembered what happened, these cases remain very much on the minds of lawmakers and law enforcement agencies throughout the country.

THE MINNESOTA PATRIOTS COUNCIL

One such incident involved a group of disgruntled right-wing individuals loosely grouped into a subunit of something called the Minnesota Patriots Council. Their particular cell was active in and around Alexandria, Minnesota in the early 1990s.[9] A group of a half-dozen or so individuals, all men, with a variety of complaints against numerous symbols of local, state, and federal authority, plotted several means of seeking revenge against those they felt were responsible for their manifold failures in life. They set out to amass various weapons, including bombs—and a potent plant-derived toxin called ricin. Highly purified ricin is, on a gram-for-gram basis, probably a hundred times more deadly than sodium cyanide, and a lot more toxic than cobra venom. They obtained the ricin by responding to an advertisement in a white-supremacist magazine published in Oregon. The ad sketched the virtues of an "assassination kit" based on ricin, which turned out to be a dozen or so castor beans (the natural source for ricin, which they were offering for one dollar per bean) plus instructions for extracting and purifying the toxin. The instructions also suggested ways to make it more easily penetrate the skin, or how to disperse it as a powder.

None of the conspirators had any background in chemistry or biology, but one of them was familiar with solvents of the type involved in extracting the ricin and volunteered to carry out its purification. They did manage to produce a white powder which they assumed to be ricin, and kept it stored in a jar in a garage. Although they talked about several potential targets for their poison and ways to deliver it, they never actually used it. As a result of dissension within the group and the defection of one of the members, the powder was turned in to the local sheriff's office and made its way to an FBI testing lab. Its identity was confirmed, leading ultimately to the arrest of four of the group. They were tried under the 1989 Biological Weapons Anti-Terrorism Act (chapter 6), which makes it a felony for U.S. citizens to produce, acquire, stockpile, or possess for use as a weapon any biological agent for which there is no discernible justification for peaceful purposes. They all received prison sentences of several years. They were the first persons to be convicted under this Act.

This case showed that it in the early 1990s it was still relatively easy to obtain at least some biological agents suitable for conversion to a biological weapon, notwithstanding that it is now (and was at the time, although this was probably not known to the individuals involved) a federal crime to do so. Ricin extraction from castor beans (which could presumably even be grown at home) is not that difficult, and detailed instructions for doing so are easily available through the Internet. The FBI recovered less than a gram of powder, which was judged to be about 5 percent pure. Still, that might have been enough to kill a hundred or more people if used effectively.

A LONE RANGER EPISODE . . .

Our final tale concerns an unusual individual named Larry Wayne Harris. Harris has an interesting pedigree: a member of an anti-government sect called the Christian Patriots, an officer in the neo-Nazi group Aryan Nations, and a trained microbiologist who carries out "research" on weaponizable pathogens in a home laboratory.[10] Raise any flags?

In May 1995, with the Aum Shinrikyo story still unfolding in the headlines, Harris ordered three vials of freeze-dried *Yersinia pestis*, the bacterium that causes bubonic and other forms of plague, from the ATCC. In order to do so, he printed up a fake letterhead on a home computer, using a made-up scientific laboratory name, his home address in Lancaster, Ohio, and an Environmental Protection Agency license number assigned to his employer, a microbial testing company in nearby Columbus.

Somewhere along the way, someone at ATCC became suspicious and alerted the CDC in Atlanta, which in turn contacted public health authorities and the police in Lancaster. The police, armed with a warrant and backed by a HazMat unit, raided his home, confiscated the *Y. pestis,* and arrested Harris.

Harris's adult life seems to consist of an endless web of fabricated stories. He claimed at various times to have been trained in biological warfare by the U.S. Army (he wasn't); to have worked as a bioweapons expert for Batelle Labs (he didn't); to have been under contract to the CIA (they deny it, as Harris said he would expect); to have trained Pentagon officials in the fine points of biological weaponry (no record); to have worked at Fort Detrick (he didn't); to have had extensive contacts with Iraqi bioweapons scientists (extremely doubtful). And so on, and on, and on.

Harris eventually pleaded guilty to a federal wire-fraud charge and received a $50 fine and eighteen months' probation. A couple of years later, while still on probation, he became involved in a complex scheme involving a machine supposedly able to kill microbes both inside and outside the body. Harris represented himself to others in the scheme as working for both the CIA and FBI as a bioweapons expert. He was expressly forbidden to do this as a condition of his probation. When Harris claimed he was going to test the machine against weapons-grade anthrax, one of the conspirators panicked and called the FBI. Harris was arrested again, this time in Las Vegas, by FBI agents accompanied by a HazMat unit and several military units specializing in bioweapons, weapons of mass destruction, and ordnance disposal.

Harris did have anthrax in his possession, but like the Aum Shinrikyo anthrax, it was a strain used for vaccinating animals, harmless to humans. The apparent ease with which he was able to circumvent then-current procedures guarding access to these agents was a wake-up call in itself to agencies presumed to be regulating them. Harris was tried again in federal court, for violation of his probation. His probation was extended by five months, and he was released.

It's hard to take Harris's shenanigans very seriously, and perhaps no one would have, except that they followed by only weeks the Aum Shinrikyo gas attacks in Tokyo and the bombing of the Murrah Federal Building in Oklahoma City. And it was quickly apparent that Aum scientists had seriously pursued plans to use bioweapons. Unquestionably, these events triggered a distinct uptick in the awareness of elected officials and policy makers, at all levels, of the threat and potential dangers of all forms of terrorism, but particularly bioterrorism. And they contributed to the burst of programs and federal spending aimed at mitigating these threats that we will examine in chapter 6.

These documented instances of the use of biological weapons (bioweapons) make clear that to date we have seen more bio*crime*, both in America and in the world at large, than what we would think of as actual bioterrorism. The perpetrators range from disgruntled individuals acting for a variety of idiosyncratic motives, through loosely defined groups often acting around religious themes. Their skills ranged from essentially zero to moderately sophisticated. The number of people they killed and injured is small by comparison with other, deadlier forms of political terrorism. And yet we have spent tens of billions of dollars to defend ourselves against bioterrorism.

So just how much of a threat is bioterrorism, in the context of the other challenges America will face in the twenty-first century? Where would the likely perpetrators of a bioterrorist attack come from? Who would they be? What can we do to protect ourselves? These are some of the questions we will address in the remainder of this book. Let's begin by first looking not at the "perps" themselves, but at the deadly agents they would likely use to terrorize us.

CHAPTER 3

AGENTS OF TERROR

THE BIOLOGICAL WEAPONS AVAILABLE TO WOULD-BE
bioterrorists today are by and large those developed during the
middle decades of the twentieth century, when most major countries were making offensive weapons for state-sponsored biological warfare. By the end of the 1960s, popular and political
opposition to biological and chemical weapons was growing, in
the United States and throughout the world. In 1969, the National
Security Council undertook a review of domestic programs for
these weapons, with input from a broad range of military and
civilian authorities, including the scientific community. Similar
reviews and discussions were taking place in other countries.

On November 25, 1969, President Richard Nixon announced
that the United States would immediately and unilaterally cease
production, and foreswear future use of, all offensive biological weapons. This might not have been an entirely altruistic
move. The United States enjoyed an overpowering superiority in
offensive nuclear weapons, and elimination of biological weapons could be seen as depriving "rogue" states of a potentially
important weapon of their own. Moreover, the U.S. military had
decided that bioweapons were probably not worth the trouble. In
addition to being difficult to produce, they were undependable
in the varying environments in which they might have to be used.
They could as easily blow back into the faces of the troops having
to use them.

After an additional two years of domestic and international discussions, in April 1972 President Nixon signed an international agreement commonly referred to as the Biological Weapons Convention (BWC), slated to come into full effect in 1975. Eventually, over 150 countries would sign it. During this interim period, the United States began systematic destruction of existing stockpiles of biological weapons and converted its major research facility at Fort Detrick to studying ways to protect American military and civilian populations against biological weapons in the future.

This latter step proved to be a wise precaution, since, as many reluctant supporters of the BWC had predicted, a number of countries continued their bioweapons programs after the 1975 effective date for this convention.[1] And to some extent the knowledge gained about how to control and contain outbreaks of a range of pathogenic organisms or their toxins (including some, such as Ebola virus and HIV, that were not even known at the time of the BWC) has already proved useful, for example in responding to the Amerithrax attacks. Unfortunately, it is entirely possible that some of the facilities still working with weaponizable pathogens, albeit for defensive purposes, may themselves be a source of deadly materials for future bioterrorists.

In the United States, the Centers for Disease Control and Prevention (CDC), a branch of the Department of Health and Human Services (HHS) centered in Atlanta, Georgia, is the primary federal agency responsible for coordinating all scientific, medical, and public health aspects of the federal response to potential and actual bioterrorism. In 1999 the CDC commissioned a detailed study of "critical biological agents" that would be most problematic in bioterrorist attacks. Those pathogens that proved to be of the greatest concern, based on factors such as lethality, ease of dissemination, contagiousness of the resulting disease, lack of preparedness of physicians and public health personnel to deal effectively with an outbreak, and ability to induce panic and social disruption, were given a Category A designation. An additional group of agents, of lesser lethality or more difficult to disseminate, were judged to be of moderate risk at present, and placed in Category B. Finally, the commission identified several emerging infectious disease agents for which there is no present history of use as bioweapons but which conceivably could be misused for such purposes in the future. These were placed in a still-evolving Category C. The CDC agents, and the diseases they cause (where defined), are listed in Table 3.1. All of these pathogens are periodically reviewed and assessed, and the list is subject to change.

In the sections that follow we take a closer look at the microbes and toxins on the CDC A list and selected agents from categories B and C.[2] While the descriptions of some of these agents can be very frightening, it should be noted that as natural elements of our environment they actually cause very little illness and death in the United States. In a sense, that is part of their potential value as weapons—there is virtually no natural immunity to these agents among the general population, and doctors have little experience in diagnosing and treating them.

ANTHRAX

Anthrax is a disease caused by a bacterium, *Bacillus anthracis*. It affects animals, mostly grazing herbivores such as sheep, goats, and cows. Humans are vulnerable to anthrax infection, but we have learned over the centuries how to avoid it, and veterinarians are skilled at keeping domestic livestock free of the disease. Prior to the early part of the twentieth century, anthrax was an occasional problem among people working with wool, hides, and other parts from animals infected with anthrax. However, fewer than 250 cases of naturally acquired anthrax in humans have been reported in the past fifty years in the United States. In less developed parts of the world, the number of annual new cases of anthrax is considerably greater. Anthrax is not contagious; there are no known cases of transmission of anthrax from one human being to another.

Anthrax is deadly: the mortality rate for untreated anthrax can range from 20 to nearly 100 percent, depending on the form of infection (see below). But what makes anthrax so deadly, and potentially useful as a bioweapon is that *B. anthracis* forms spores. Most bacteria, when they run out of food, simply starve to death. A few bacteria, however, are able to enter a state of suspended animation—to convert to bacterial spores. Spores do not carry out any metabolism, do not need water, and are very resistant to heat and many toxic chemicals. Properly prepared, they are hard, dry particles that can be dispersed in air. When they land on a surface possessing moisture and nutrients—human skin or lungs, for example—they rapidly revert from spores to normal bacterial cells in a process called germination. Ungerminated spores can survive in soil or on inert surfaces for several decades.

In *inhalation anthrax,* spores settle into the lungs, and within hours, actively dividing bacteria migrate through lymph and blood to other parts of the body. It doesn't take long before anthrax bacteria have spread everywhere. They then release toxin molecules which can seriously damage a wide range of tissues and organs. Depending on the number and quality of spores inhaled, symptoms of disease can appear anywhere between a few days and a week or two. Initial symptoms are similar to those of the flu and may be misdiagnosed. But once it sets in, the disease accelerates very rapidly, with high fevers, vomiting, and diarrhea. The body is quickly overcome by bacterial toxic shock, which can be followed by coma and death.

A hundred years ago, mortality in untreated inhalation anthrax could easily approach 100 percent. But the Amerithrax incident, if it can be thought of as having an upside, suggests that rapid diagnosis and aggressive treatment with antibiotics[3] can reduce mortality to somewhere in the 50 percent range. Part of what makes anthrax so hard to manage clinically is that even after the bacteria have been killed, the toxins they have released (Box 3.1), which are not affected by the antibiotics, remain in the system for at least a day or two, and continue to cause damage.

Spores settling on healthy skin are unlikely to cause a problem, but they can enter the body through cuts or abrasions and cause *cutaneous anthrax.* Once inside, some spores will germinate locally and cause redness and itching that can develop into local skin ulcers; these eventually turn a deep, coal-like black color, whence the name *anthrax* (from the Greek word for coal). Many spores will spread to other parts of the body, germinating as they go. However, death from this form of anthrax, even untreated, rarely exceeds 25 percent. With prompt antibiotic treatment, mortality is rare. None of the Amerithrax victims with the cutaneous form of anthrax died.

BOX 3.1

TOXINS RELEASED BY ANTHRAX BACTERIA

Edema toxin	Causes massive leakage of water from blood vessels into tissue spaces. Inhibits white blood cells involved in fighting infections.
Lethal toxin	Causes massive inflammation of body tissues, leading to systemic shock.

SMALLPOX

Today it is hard to imagine that smallpox was once one of the deadliest diseases on this planet, exceeding even the plague in the cumulative number of people killed throughout history. In the twentieth century alone, it is estimated half a billion people may have died of smallpox. When contracted through the lungs, by breathing in air into which an infected person had sneezed or coughed, it routinely killed a third of unvaccinated individuals well into the twentieth century, and left the rest badly disfigured for life. Smallpox can also be spread by person-to-person physical contact, although the resultant disease is less fatal.

Smallpox in humans is caused by a virus, *Variola major*.[4] It is one of a relative handful of human pathogens that has no known animal or insect reservoir (a host in which it can reproduce without causing disease, or at least death.) In its present form, it appears to be entirely dependent on human beings for its propagation and survival.

In the course of a *V. major* infection, viruses settle into airway tissues and are swept along into regional lymph nodes. After a week to ten days, various combinations of fever, chills, and achiness appear. When the virus reaches the skin (from the inside out, as it were), a rash appears, followed by the formation of multiple, closely packed blisters on all parts of the body, but particularly the face and neck. This may not occur until ten to seventeen days after initial infection. These blisters also form in the mouth and throat, where they break easily, dumping their viral load into the saliva. This aids in the further spread of the virus into the general population through coughing and sneezing. *V. major* was probably endemic in the human species for thousands of years. From the eighteenth century—in some parts of the world, even earlier—it was kept under partial control through the natural immunity of recovered victims and by various forms of active vaccination. It was the absence of both natural and vaccine-induced immunity that made Dark Winter possible.

V. major is the first (and so far only) disease-causing microbe to be purged from the human species, by a worldwide immunization campaign launched by the World Health Organization in 1967. The fact that *V. major* could not retreat into an animal reservoir during this campaign was probably a factor in its eradication. Today *V. major* officially exists only as frozen stockpiles at the CDC in Atlanta and in a former biological warfare research center near Novosibirsk, in Russia.

Immunizations for smallpox over the years have never been carried out with *V. major*—it is too deadly—but rather with a closely related orthopoxvirus called vaccinia, which causes cowpox in cattle. A relative of *V. major*, vaccinia is the virus used by Edward Jenner at the end of the eighteenth century to become the first person to induce immunity to a disease in humans, and is the origin of the term vaccination.

Vaccinia is injected in a fully viable form. In humans it induces a mild local reaction at the site of injection that usually resolves in seven to ten days. Protection from subsequent infection by *V. major* after vaccinia immunization is excellent, but with even the least pathogenic forms of this virus, about 1.6 cases per million immunizations progressed from mild reaction to more serious disease and occasional deaths, which is why vaccination for smallpox in the United States was halted in 1972.

Smallpox is on the CDC Category A list because of its high mortality rate, and because *V. major* spreads very efficiently as an aerosol. The virus is relatively stable, and as a viral disease, smallpox is essentially untreatable.[5] As with anthrax, there is enough residual public awareness of the deadliness of smallpox that news of its spread in a terrorist attack would likely generate considerable panic and social disruption. Since for the past thirty years or so almost no one in this country has been vaccinated against small-pox, the U.S. population is highly vulnerable to this disease.

PLAGUE

References to the plague in human history date back to at least 500 BC, although we don't really know if that plague was the same as the great pandemics that swept Europe and Asia in the Middle Ages, which are what we recognize as plague today. There were several major and many minor pandemics in Europe in the four-teenth through eighteenth centuries. They were deadly. Although we have no precise figures, somewhere between 100,000,000 and 200,000,000 people are estimated to have died.

The non-spore-forming bacterium *Yersinia pestis* has been asso-ciated with the European outbreaks, and *Y. pestis* DNA has actually been extracted from dental remains found in graves of persons dying from the plague. We still see occasional incidents of *Y. pestis* plague, with several thousand new cases arising annually through-out the world. There have been about 400 cases in the United States since 1950, mostly in the Southwest.

Y. pestis can cause several types of disease in humans, depend-ing on how the infection is acquired. *Bubonic plague* results when a human is bitten by an insect, usually a flea, carrying *Y. pestis* acquired from previously biting an infected animal. In urban areas, the most common animal carriers are rats and squirrels and the occasional house cat. *Y. pestis* can also jump from animal fleas into fleas feeding on humans, which greatly aids human-human spread of the disease. In the Middle Ages, most people had fleas in abun-dance. Prior to the introduction of antibiotics in the 1940s and '50s, fatality rates of 50 percent or more were not uncommon.

A few days after infection, the typical symptoms of a microbial infection set in: fever, chills, and general achiness, by-products of activation of the immune system. As the bacteria continue to replicate inside regional lymph nodes, the nodes become greatly enlarged (called buboes) and very tender. They can grow as much as four inches across.

If the bacteria spill out of the lymph nodes and enter the general blood circulation, numerous other tissue compartments become involved, and the infection is even more lethal (septicemic plague). Blood vessels are destroyed, resulting in gangrene in the extremities; this is probably the origin of the term "Black Death" for plague. Prolonged infection can also trigger shock, a common cause of plague death. If the bacteria invade the lungs (secondary pneumonic plague), the infection is almost always fatal, and the bacteria spread more readily from person to person through sneezing and coughing.

The second major form of plague, primary pneumonic plague, is particularly deadly. It occurs when *Y. pestis* is taken in directly through the respiratory system as opposed to an insect bite. Untreated mortality rates approach 100 percent. Symptoms set in within a day or two after inhalation of infectious *Y. pestis*, and are initially indistinguishable from other forms of aggressive pneumonia. Experience in diagnosing and treating pneumonic plague is very limited in the United States. Moreover, many currently used antibiotics have never really been tested against *Y. pestis* in humans.

There have been no documented attempts to use *Y. pestis* as a bioterrorism agent. However, in 1995, a microbiologist was arrested for fraudulently obtaining large amounts of plague bacteria with no obvious legitimate scientific purpose. And in 2004, a respected physician-scientist at Texas Tech University was sentenced to two years in prison for grossly mishandling and illegally shipping to Tanzania vials containing infectious *Y. pestis*—on a commercial airliner, no less! No connection with bioterrorism was alleged or proved.

BOTULISM

Botulism is caused by a protein toxin released by several strains of bacteria of the genus *Clostridium,* including the eponymous *C. botulinum.* This toxin is the most lethal biological poison known—100,000 times more poisonous than sarin gas.

Natural infection by *C. botulinum* can occur by eating contaminated food, usually vegetables, though the name botulinum in fact derives from the Latin for sausage (*botulus*), a common food contaminated by *C. botulinum* in former times. The poisonous effect of food contaminated with this bacterium is due entirely to the toxin, which it readily secretes into its surroundings. *C. botulinum* can also enter through wounds, for example in the foot, when walking in soil harboring this bacterium. The toxin itself will not penetrate unbroken skin.

Botulinum toxin is a neurotoxin that prevents the brain from telling muscles when to contract. Within twelve to seventy-two hours after ingestion (depending on dose), individuals begin experiencing muscular weakness, and have difficulty seeing, speaking, and swallowing. Treatment almost always requires extended stays in intensive care units. While there may be some initial sense of giddiness, the brain and associated mental functions are not impaired. But within hours the muscles, including those that control breathing, just cannot do work. Death usually comes quickly from respiratory failure.

There are seven genetically distinct strains of *C. botulinum*, each producing a slightly different neurotoxin, given the designations A through F. Almost all cases of botulism in humans are caused by toxin types A, B, and E.

Intact *C. botulinum* bacteria were fed by the Japanese to prisoners during their occupation of Manchuria in World War II. The results, as far as they are known, were uniformly lethal. The Aum Shinrikyo cult in Japan, as we have seen, attempted to carry out attacks in Tokyo in the 1990s using botulinum toxin, but were unsuccessful. Research on aerosolized botulinum toxin as a possible biological weapon was carried out by the United States and other countries over the years, but the toxin was never used. United Nations inspectors determined during the 1990s that Iraq had prepared about 5,000 gallons of concentrated toxin, some of which was reported by inspectors to have been loaded onto missiles, ready for use.

TULAREMIA

Tularemia is caused by a bacterium, *Francisella tularensis*, named for its discoverer, Edward Francis, and the place of its discovery, in 1911, in Tulare County, California. There are two major strains of *F. tularensis*, Type A and Type B. Type B is the most virulent, being among the most infectious pathogens known. It had been associated with a variety of plague-like diseases in animals such as deer-fly fever, rabbit fever, and tick fever, among others, all of which are now grouped as various forms of tularemia. Humans can be infected by contact with *F. tularensis* in the wild, although such incidents are relatively rare (only about 200 cases per year in the United States), and humans do not readily transmit the resulting infection to others.

The most serious incidents of tularemia in humans come from inhalation of bacteria, often through handling of contaminated hay or other grains. "Inhalation tularemia" requires only a few bacteria to cause serious illness, whereas infection through other routes usually requires exposure to millions of bacteria. Because of the low incidence of inhalation tularemia in the United States, a large outbreak of this disease in a concentrated area might take a while to diagnose properly, but would lead to an immediate suspicion of bioterrorism.

Tularemia was investigated by several countries between 1930 and 1970 as a potential biological warfare agent. It is thought to have been genetically altered to an even more deadly form by the Soviets in 1986, which would make it the first such agent to be so modified. The bacterium can be concentrated into a paste, which can be freeze-dried and then milled into a fine powder suitable for distribution through the air.

HEMORRHAGIC FEVER VIRUSES

Hemorrhagic fever viruses (HFVs) are, if not the most deadly of human pathogenic microbes, certainly the most dramatic. The CDC has designated nine HFVs as potential bioterrorism agents (Table 3.1). We focus here on only two of these, the Ebola and Marburg viruses. Both of these viruses are fairly recent additions to the repertoire of human pathogens, and not much is known about their interaction with their human hosts. There have only been a dozen or so outbreaks of these viruses since their discoveries in 1967 (Marburg) and 1976 (Ebola).

We do not know what animals serve as reservoirs for these viruses; bats are one likely candidate. Several nonhuman

TABLE 3.1   CDC Classification of Potential Bioterror Agents

Category	Agent	Disease
A	B. anthracis	Anthrax
	V. major	Smallpox
	Y. pestis	Plague
	F. tularensis	Tularemia
	C. botulinum (neurotoxin)	Botulism
	Ebola, Marburg viruses[1]	Hemorrhagic fever
B	Brucella sp.	Brucellosis
	E. coli O157:H7	Food poisoning
	B. mallei	Glanders
	B. pseudomallei	Meliodosis
	C. perfringens ε–toxin	
	C. psittaci	Psittacosis
	C. burnetii	Q fever
	R. communis	Ricin poisoning
	Staphylococcus sp. enterotoxin B	Food poisoning
	R. prowazekii	Typhus
	Alphaviruses (various)	Viral encephalitis
	V. cholera	Cholera
C	Nipah virus	
	Hantavirus	
	MDR[2] M. tuberculosis	Drug-resistant TB
	Yellow fever virus	Yellow fever
	Avian influenza virus H5N1	Pandemic influenza

[1]Other hemorrhagic fever viruses in Category A include Lassa fever virus; four
New World arenaviruses; Rift Valley Fever virus; Omsk hemorrhagic virus;
Kyasanur Disease virus.
[2]MDR = multi-drug resistant.

primates, such as rhesus monkeys and macaques, are fully susceptible to Marburg and Ebola infection and would be unlikely
reservoirs—their mortality rate is close to 100 percent. Once one
person has been infected, transmission to others occurs through
contact with fluids or tissues from previously infected individuals. Although the disease is contagious, infected individuals are

usually quickly isolated and do not live long, so widespread epidemics are rare.

Ebola and Marburg viruses would certainly fulfill the CDC requirement that an agent have the potential to cause "public panic and social disruption." Through books and films in the last dozen years, as well as regular media coverage of outbreaks, these viruses may have, along with anthrax, the highest public profile of the Category A agents. Because of the high mortality rate, the fear factor may be even greater than for anthrax. As of mid-2005, 1,848 cases of HFV in humans caused by Ebola had been reported to the WHO, with 1,287 fatalities (69.9%); 354 cases of Marburg HFV had been reported, with 288 deaths (81.3%). Almost all of these cases arose in Africa. Many have been traced to transmission through unclean clinical syringes, a common problem in rural Africa; the resulting mortality in these cases was 100 percent. We might hope mortality would be somewhat lower in industrialized countries, but make no mistake: these viruses are far and away the most lethal pathogens on the CDC's A list.

The Soviet Union was alleged to have produced aerosolized versions of HFVs, including Ebola and Marburg viruses, for biological warfare. These were never used, and we have no idea how effective aerosolized HFVs would be. Aerosolized HFVs are relatively stable and cause disease in nonhuman primates, and it is presumed they would do the same in humans.

The popular depiction of humans being literally melted away from the inside out draws on a good deal of dramatic license, but these are undeniably ghastly diseases. Blood vessels as well as blood cells are rapidly destroyed once the virus begins to spread in the body, causing massive internal bleeding. But organs such as the liver and kidneys are also severely damaged. Symptoms of hemorrhagic fever include the usual early signs of any microbial infection, but these are quickly followed by widespread body rashes and blood spots in the skin, blood seepage from various orifices, convulsions, delirium, and a rapid descent into shock and coma. There are no anti-viral drugs effective against Marburg and Ebola. Both viruses severely dysregulate the human immune system, leaving most victims with no innate resistance. For the few people who survive, there is a long period of impairment of numerous body functions.

Attempts to produce an effective vaccine against Ebola and Marburg had been generally unsuccessful until 2005, when a research group centered in Canada developed a DNA-based vaccine that is very potent against *both* Ebola and Marburg. Just one injection protected monkeys from infection by either virus. It is possible that with further work this vaccine could also be made effective for other A-list HFVs. More work needs to be done before human trials can begin, but this vaccine looks extremely promising.

SELECTIONS FROM THE CDC B AND C LISTS

RICIN TOXIN

Ricin toxin (usually just called ricin) is a protein extracted from castor beans harvested from the plant *Ricinus communis* and pressed to obtain castor oil. In purified form, ricin can be extremely poisonous; a dozen or so beans contain enough toxin to kill an adult human. It works by blocking protein synthesis inside cells, causing them (and eventually the tissues and organs they make up) to fail. It can be prepared as a dry powder or as an aqueous liquid. If inhaled as a powder, symptoms set in within a few hours, including chest pain, difficulty in breathing, coughing, and nausea. The lungs gradually fill with water, blood pressure drops, and at high enough dose the victim dies of a combination of shock and respiratory failure. Swallowing ricin results in vomiting and diarrhea, leading to rapid dehydration. Hallucinations, seizures, and other neurological problems may occur. With a sufficiently large dose, liver and kidney function may cease, leading to death.

Part of the attraction of ricin as a bioterror agent is that there is no diagnostic test for it, once ingested, and no antidote. Most countries developing bioweapons experimented with ricin, but no weapons appear to have been manufactured or used. Ricin was used in the assassination of a dissident Bulgarian writer, Georgi Markov, in London in 1978 (Box 3.2).

BOX 3.2

A BIOTERRORIST ASSASSINATION

Georgi Markov was a Bulgarian writer and dissident who fled
Sofia for England in 1969. Working for Radio Free Europe and
the BBC, he wrote stinging articles about the Bulgarian regime.
With KGB help, it was decided to assassinate him. At a bus stop
near Waterloo Bridge in London, in September 1978, an agent
stabbed him with an umbrella tip which was actually a gas-driven
gun. Markov died three days later. At autopsy, a 1.5-mm pellet was
found, but its significance was not clear. A similar pellet found in
another assassination attempt in a Paris Metro (the victim lived)
proved to contain highly pure ricin. An examination of Bulgarian
records after 1990 and other evidence suggested a Dane of Italian
origin was the likely assassin. He was never formally charged with
the crime for lack of evidence. The statute of limitations will run
out in 2008, at which time we may learn more.

STAPHYLOCOCCUS ENTEROTOXIN B

Enterotoxin B is a protein toxin secreted by various strains of the
common food bacterium *Staphylococcus aureus*. It works by hyper-
activating the immune system, triggering excessive release of
numerous chemicals which in moderate doses help regulate the
immune system's response to many microbes but which, in great
excess and duration of action, can cause severe damage.

Enterotoxin B is a classic and potent agent of food poisoning, causing intense intestinal cramps, nausea, vomiting, and
diarrhea within a few hours of ingesting staphylococcus-tainted
food. In amounts normally associated with this mode of intake,
the symptoms are self-limiting and disappear within twenty-four
hours.

Enterotoxin B has been studied as a potential bioweapon
because in its highly purified form it can be made into a stable,
easily aerosolized powder which, when inhaled, can lead to profound incapacitation through acceleration and intensification of
the symptoms described above. With concentrated intake of pure
enterotoxin, the symptoms are no longer self-limiting, the loss of

fluids can become crippling, and the result may be toxic shock syndrome and death. There is no vaccine to enterotoxin B, and no antidote, since under normal conditions of food poisoning there is no danger to health.

Q FEVER

Q fever is caused by a rickettsia-like bacterium, *Coxiella burnetii.* It is found in many domesticated animals but does not cause disease. They thus serve as a reservoir for this bacterium, which can cause disease in humans. Natural infections in humans occur mostly in people working around farmyards or slaughterhouses, where waste products from infected animals become ground down, dried, and eventually airborne. The bacteria are very stable and can be inhaled by animals or humans. Passage of *C. burnetii* between humans is very rare, and Q fever is thus not considered contagious.

Natural infections with *C. burnetii* can be acute or chronic. Acute infections are characterized by high fever, lethargy, vomiting, diarrhea, and weight loss and can last several weeks. If not treated promptly, acute Q fever may also progress to crippling headaches, liver problems, and speech and hearing difficulties. Most cases respond well to antibiotic treatment, and mortality is rare.

Chronic Q fever may have the above symptoms in varying degrees for six months to many years, and patients with this form of the disease usually develop heart problems as well. Chronic infections are very difficult to treat, and mortality can approach 60 percent. Because of the ability of *C. burnetii* to become airborne and its general hardiness, it has been considered a good candidate for use in bioterrorism. Aum Shinrikyo is known to have been interested in this pathogen, although there is no evidence they succeeded in making useful preparations of it.

HANTAVIRUSES

Hantaviruses are an example of an agent that has never been used as a bioweapon, for terrorist purposes or otherwise, but which the CDC is definitely keeping an eye on for one simple reason. In our limited experience so far with this virus, the death rate appears to be between 30 and 40 percent. In 2005, for example, there were only thirty-four confirmed cases in the United States, but ten of these resulted in death. The natural reservoir for hantaviruses is various rodents, such as field mice. Initial symptoms are hard to distinguish from other microbial infections, but within a few days dry cough, diarrhea, and dizziness, together with a drop in blood pressure and a racing pulse, suggest something out of the ordinary. Detailed examination of the blood and a complete cardiopulmonary workup provide additional indications of hantavirus.

WHAT ABOUT HIV?

The question is often asked: "Why wouldn't bioterrorists use HIV as a weapon? Why isn't it on the A list?" Unquestionably, the release of HIV over a large metropolitan area could generate a maximum fear effect. And as we know all too well, all but a tiny handful of us are defenseless against HIV, with no vaccine on the immediate horizon. So the fear factor probably extends to would-be terrorists themselves, both domestic and foreign. They may be extremely reluctant even to get into the same room with HIV. Another factor is that the incubation period with HIV, before frank (full-blown) AIDS sets in, is six to ten years. Suspiciously large numbers of new cases would likely not be apparent for several years at a minimum. The immediate public relations sensation so craved by terrorists would be lost.

Still, the overall psychological impact on affected populations could be enormous. In the end, the main thing preventing use of HIV is that this is an exceptionally fragile virus. Exposure to anything other than a warm, wet human body disables it within a matter of hours. Aerosolization would almost certainly cripple it. Laboratories working with HIV must take enormous care to keep their strains viable. It is, in fact, a poor candidate for even the CDC's C list.

CHAPTER 4

GENETICALLY MODIFIED
PATHOGENS

IT HAS BEEN SAID MORE THAN ONCE THAT MILITARY FORCES are usually trained to fight the last war. Most of the pathogens on the CDC lists are in that mode—pathogens we already know from having explored their use, and defenses against them, in previous times. All of them exist in nature, and have changed very little since humans began studying them. The threat of bioterrorism based on these pathogens has stirred intense research into the development of new vaccines and drugs to defend against them, and we have made remarkable progress.

In fact, it is likely that within not too many years, we will have effectively neutralized most if not all of the CDC agents as potential bioterror weapons. But we should not assume that these are the only biological weapons that might be used against us. For some time now, scientists have been asking what the next generation of bioweapons might look like and how we can prepare ourselves to defend against them.

Do we really have to worry about a "next generation" of biological weapons? After all, didn't the Biological Weapons Convention of 1972 effectively shut down research into offensive biological weapons? Maybe. We think our own government has stopped such research, and we hope that others have done the same. But we don't really know. And how do we define research into biological weapons, particularly if we're trying to think outside the box of today's arsenal?

Concern about new bioweapons systems arises from advances in technology, in a field known as molecular biology, that permit genetic modification of existing pathogens, or even creation of new pathogens from scratch. Genetically modified pathogens could be made resistant to existing antibiotics or vaccines. They could be made more stable, or more easily weaponized. They could be equipped with new toxins or other molecules that make them even more deadly. Extension of these same technologies could even be used to recreate ancient pathogens that no longer exist in nature, or to create new ones never before seen.

All of these technologies depend to a large extent on the revolution in molecular biology known as recombinant DNA. This refers to the ability to cut a piece of DNA out of one source and splice it into DNA from a different source—or more to the point, to take a gene from one living thing and put it into another.

The recombinant DNA revolution began in 1973 when scientists cut out a gene from toad DNA and inserted it into the DNA from a benign form of *E. coli* bacteria that lives in the human gut. The bacteria promptly began making the corresponding toad protein. Such a thing had never been done before. It was clear to anyone who cared to think about it that biology was going to change, big time. Other scientists began recombining DNA taken from a virus called SV-40, which causes cancer in monkeys. The resulting recombinant genomes[1] were, in effect, totally new life forms—never before seen among organisms that had evolved naturally over eons of time.

The problem was that *E. coli* lives, among other places, in the human gut. And SV40 can infect humans. How would these new life forms behave? How would we interact with them, and they with us? There was no evidence that such experiments were in fact dangerous, but the suggestion was made that scientists might want to suspend further experimentation until everyone could get together for a talk.

A meeting to discuss the implications and possible risks of recombining the DNA of living organisms took place at the Asilomar Conference Center near Monterey, California, in 1974. All of the major laboratories working with recombinant DNA were invited, along with representatives of the federal agencies funding such research. To forestall possible charges of a scientific elite meeting behind closed doors to decide the biological fate of humanity, the press was invited to attend as well. Many issues, some highly emotional as well as scientific, were aired in the formal sessions, but to an even larger extent over meals, in the common rooms where people gathered in the evenings, and in strolls along the spectacular adjoining beaches. Would the creation of recombinant genomes interfere with normal evolutionary processes? Do human beings have the right to reach into nature and create new life forms?

But a few, even at the time of the Asilomar Conference, looked ahead to an even less comforting possibility—that someday, someone could decide to use recombinant DNA technology to modify human pathogens in ways that would make them more of a threat than they already were. And indeed, such genetic alterations were not long in coming. There is compelling evidence that the Soviet Union pursued this line of research quite extensively in the 1980s and maybe even into the 1990s. We learned a great deal about their various bioweapons programs from Soviet defectors who had worked on them.[2] These were for the most part well-educated, smart, technically competent scientists who have since contributed a good deal to our own biodefense programs.

Their activities, organized under a government apparatus known as Biopreparat, are thought to have involved dozens of laboratories, with hundreds if not thousands of scientists and technicians. At one time or another, they produced genetically enhanced versions of most of the pathogens on the CDC's A and B lists. *F. tularensis,* perhaps the least dreaded of the A-list pathogens, was one of the first to be engineered, in 1983. They had at least one strain of anthrax resistant to the antibiotic Cipro. They claimed to have created a strain of *Y. pestis* resistant to ten different antibiotics. In addition to antibiotic and vaccine resistance, they also allegedly produced germs capable of crippling a person's immune system or turning the immune system against the nervous system (Box 4.1). Most of these worked well in animals but as far as we know were never tested against humans.[3] Plans were even floated to create viruses that carried not disease, but a whopping dose of Prozac!

BOX 4.1

PARTIAL LIST OF GENES USED TO MODIFY PATHOGENS

Genes to induce disease

Encephalitis genes
Myelin basic protein gene
Ebola vp24 gene
Marburg NP gene

Genes to produce physiological alterations

β-Endorphin gene
Angiogenin gene
Insulin gene
Thymosin gene

Genes to perturb immune system

Interferon genes (α, β, γ)
TNF α, β genes
IL-2 gene

It would take a relatively short time to produce new variants of CDC A-list pathogens in first-rate molecular biology labs like those the Soviets had. But if such a new germ were used in a bioterrorist attack, it could take a good deal longer for responding authorities to figure out, first of all, that such a change had in fact been made, and then to devise and test a strategy to neutralize it.

SYNTHETIC BIOLOGY

Our concerns about possible threats posed by these new technologies have taken a surprising new turn with the emergence, through the coalescing of various DNA manipulation techniques, of the new field of synthetic biology. Classical recombinant DNA experiments focus largely on the insertion of discrete pieces of DNA, usually a gene or a portion of a gene, from one organism into the genome of another. In synthetic biology, entire portions of the genome of one organism are recombined with varying portions of another to create new hybrids or, most impressive of all, to create an entire organismal genome from scratch, starting with the basic building blocks of DNA called nucleotides. Let's look at three examples of what can be done using these newer techniques.

POLIOVIRUS

Polio (poliomyelitis) is a paralytic disease caused by a virus—called simply poliovirus—that attacks and destroys nerve cells in the spinal cord. The disease no longer occurs naturally in the United States. Through the early 1950s, it was common to see about 20,000 cases annually, but the development of polio vaccines in the mid-1950s reduced that dramatically. The last natural case of polio in the United States occurred in 1979. Since 1980, the only cases to appear have been picked up in other countries, or resulted from the oral polio vaccine, which was discontinued in 2000. The WHO declared Europe polio-free in 2006. Poliovirus, like the smallpox virus, exists only in humans, with no animal reservoir, so the stated WHO goal of complete global eradication of the poliovirus in the coming decade, through an intense immunization campaign, seems realistic.

But now we're not so sure.

Poliovirus is a picornavirus: it has a tiny (pico) RNA (rna) genome. Only 7,741 nucleotides are required to make its entire genome; the smallpox virus, by comparison, takes more than 185,000 nucleotides. In 2002, scientists at the State University of New York at Stony Brook reported making a complete, infectious, pathogenic poliovirus, essentially in a test tube, from its chemical building blocks—nucleotides—which are not on anyone's select list.

The precise RNA nucleotide sequence had long been known—poliovirus was one of the most intensely studied of human viruses. Using the published sequence of the virus, the researchers enlisted a private company, via the Internet, to manufacture individual small segments of the poliovirus genome, and then stitched these together to get a complete viral genome. When this synthetic genome was placed into human cells in culture, it directed the production of intact poliovirus, which was shed into the culture dishes. When these "frankenviruses" were injected into mice, they caused paralysis and death indistinguishable from naturally occurring poliomyelitis. This was the first time that a virus capable of infecting living cells was created from chemical materials entirely in a laboratory. But it wouldn't be the last.

φX174

In 2003, a group of scientists at a research institute in Maryland carried out a similar set of experiments with a virus that infects not animals but bacteria. Such viruses are called bacteriophage (φ). The one generated by this group, called φX174, is even tinier than poliovirus (5,386 nucleotides). It has been used in numerous famous experiments over the years; it was the first genome to be duplicated from another genome copy entirely in a test tube (1967), and the first genome to be sequenced in its entirety (1978). The basic strategy used to make the φX174 genome from scratch was similar to the poliovirus work, but the efficiency was ten times greater. The team is moving forward to synthesize a complete bacterial genome—a chromosome—some 300,000 nucleotides in size. If they accomplish this, and can succeed in getting their artificial chromosome into a cell, they will have a chance to be the first to actually create life itself, beginning with raw chemicals, in a laboratory.[4]

INFLUENZA A VIRUS

The flu virus is well known to just about everyone. Like poliovirus, the flu virus is a small RNA virus, with only eight genes (versus our 30,000 or so). For the most part, it causes annoying but, for reasonably healthy people, nonthreatening flu each year. However, it has an ability (like the AIDS virus, HIV) to rearrange itself from time to time, confounding our immune systems. The immunity we build up one year may not recognize the form of the virus that comes our way the next year. And sometimes the flu virus can generate variants that are more than just annoying; they can be very deadly, as with the 1918 influenza virus that caused a worldwide flu pandemic, killing up to a hundred million people (chapter 5).

The 1918 form of the flu virus has not been seen since the pandemic it triggered subsided, and is presumed to no longer exist in nature. But its tiny RNA genome, recovered from preserved autopsy materials and from a corpse buried since 1918 in the Alaska permafrost, has brought it back. In 2005, a group of researchers from four different universities and institutes published a rather startling version of the "from scratch" experiments in the journal *Science*. Using the sequence for the 1918 *H. influenzae* genome, they built up progressively larger subsequences and finally stitched them together to make an intact viral genome.

When viruses produced from this genome were used to infect human cells in culture, they grew at a rate fifty times faster than normal flu strains. When used to infect mice, the reconstructed 1918 strain not only reproduced itself at an accelerated rate but proved to be a hundred times more lethal than normal flu strains. And when injected into macaque monkeys, it caused a rampant disease that began killing the monkeys in a matter of days. The remaining animals were euthanized because of extreme suffering.

Scientists are now studying exactly what happened in these monkeys, in an attempt to understand the sequence of events in humans attacked by this and similar viruses. It appears that the incredible damage wrought by this virus came about in part because it triggered a huge overreaction by the host immune system, and may be yet another expression of what is known as *immunopathology*, where the damage done during an infection is caused as much by the immune system that is supposed to defend us as by the invading microorganism.[5]

One major concern associated with synthetic biology is that much of the synthesizing of DNA or RNA sequences used in the creation of functional genomes is now done by commercial companies. Originally a time-consuming laboratory procedure requiring constant oversight by experienced scientists, much of this work is now carried out very rapidly by fully automated machines. These machines are available in many major universities as well as in private biotechnology firms. In principle, anyone with enough money has access to the latter, and the former are, at least at present, under rather loose control.

In the summer of 2004, several hundred scientists carrying out research in synthetic biology gathered at MIT for the First International Conference on Synthetic Biology. This is a de facto way of recognizing the existence of a discrete line of scientific inquiry as a new scientific field. The meeting was called to promote interactions among researchers in the new field. Program topics ranged over a wide spectrum of cutting-edge technical and conceptual advances.

The possible abuse, by terrorists or other biological "hackers," of the tools used by synthetic biologists was not itself a programmed topic of discussion, but it worked its way into several formal presentations and many sidebar discussions throughout the meeting. The clear requirements in the 2002 Bioterrorism Act for close regulation of certain pathogens and for licensing of users and suppliers of select agents were well known to most of the attendees. There was considerable informal discussion about the extent to which aspects of synthetic biology relating to the synthesis or re-engineering of potential pathogens could or should be regulated.

A second, less formal gathering of synthetic biologists took place the following summer in Berkeley, California, but this time discussions of legal and ethical issues generated by the new field made their way onto the formal program. Ghosts of Asilomar Past must have wafted throughout the symposium. Speakers sometimes felt it necessary to phrase their remarks in terms of the "realpolitik" of post-9/11 America.

The Second International Conference on Synthetic Biology was held at UC Berkeley in May 2006. Again the main focus was on technical and scientific exchanges, but this time the legal and technical aspects of the field were recognized in a session that was also a formal part of the program. Outside the lecture halls, intense informal discussions about regulatory issues, particularly among senior scientists, were a hallmark of the meeting. After the conference was over, the organizers prepared a "white paper" on security and legal issues, with recommendations for how the field should proceed.[6] Some of the proposals are shown in Box 4.2. This paper was published on the Web for comment by the larger scientific and general public, after which a formal position paper will be issued.

BOX 4.2

PROPOSALS FOR INCREASING THE SAFETY OF
SYNTHETIC BIOLOGY RESEARCH

1. Improve the ability of current software used by companies that generate DNA segments to spot sequences related to known "select agent" pathogens.
2. Require all commercial DNA companies to scrutinize each order for DNA with respect to source of the order, credentials of the orderer, etc. Researchers should boycott companies that do not undertake such scrutiny.
3. Create interdisciplinary groups to address challenges arising from advances in synthetic biology, especially legal and security issues.
4. Support ongoing dialogues with all stakeholders to develop and analyze how best to regulate synthetic biology research without crippling it.

Environmental groups are already arming themselves to oppose synthetic biology entirely, or at least to have it regulated extremely tightly.[7] At the UC Berkeley Conference, a letter signed by thirty-five environmental groups, trade unions, and bioethicists implored the attendees to work for strict national and international controls on their research. Aside from concerns they may share about bioterrorism, they are worried about the effects that genetically altered microbes might have should they escape or be accidentally released back into nature. So are many scientists. The Third International Conference is slated for 2007, in Zurich. It is likely that in the future, environmental groups will be invited to sit at the table.

COUNTERMEASURES

By the late 1990s, both the CIA and the Pentagon were becoming seriously concerned that the United States could be vulnerable to the types of bioweapons the Soviet Union was known to have been working on after the 1972 Biological Weapons Convention finally went into effect in 1975, including genetically modified pathogens. In the late 1990s, the CIA created a special project called Clear Vision to review everything that was known about Soviet ventures into modified pathogens and their delivery systems. Clear Vision would also gather and evaluate all intelligence leads about work on genetically modified pathogens in other countries considered potentially hostile to the United States.

Some scientists working in Clear Vision tried their hand at designing virtual (computer-generated) pathogens, to get a sense of what kinds of genes might be moved into and out of a pathogen to make it more deadly, or resistant to vaccines and drugs. The Soviets, they believed, had spliced a gene for diphtheria toxin into a plague genome, creating a genuine monster of a pathogen. They were also thought to have stitched in genes that blocked a victim's immune system from responding, depriving the victim of any first-line defense and greatly speeding up the action of the altered pathogen. Clear Vision scientists wanted to find ways to rapidly detect, identify, and counter these mutant killers. In order to do so, they wanted to build live models of them in the laboratory.

Many officials and even some scientists outside the project began to worry that engaging in this kind of research could move the United States itself perilously close to violation of the 1975 Convention, which forbids signatory states to "develop, produce, stockpile, or otherwise acquire or retain microbial or biological agents, or toxins, that have no justification for prophylactic, protective or other peaceful purposes." By 2001 the program had ground to a halt over such concerns.

But the biodefense concerns embodied in Clear Vision were carried forward in new form when the Department of Homeland Security created the National Biodefense Analysis and Countermeasures Center (NBACC), located at Fort Detrick, Maryland, in 2005. The NBACC is composed of two parts. The National Bioforensic Analysis Center will act as the lead government agency in analyzing materials recovered from any bioterrorism attack site. For the most part, this would consist of rapid analysis of the DNA of recovered pathogens or chemical analysis of any toxin, either of which could determine whether the material had been altered and provide "fingerprints" pointing to possible origins of the materials.

The Biological Threat Characterization Center will carry forward much of the work of Clear Vision. In what is being called "science-based risk assessment," scientists at the NBACC will have as a principal goal anticipating the development of previously unknown bioweapons and devising methods to counter them. They will conduct analysis of the current scientific literature in the area of genetic engineering, as well as intelligence reports about scientific centers in countries estimated to have the potential (and the political will) to engage in the production of new bioweapons.

The NBACC will also be looking at technologies to rapidly detect genetically altered pathogens, faster ways of developing drugs and vaccines that might be used against them, and more effective pathogen decontamination procedures for dealing with the aftermath of an attack with such pathogens.

Although assurances have been provided that safeguards are in place to be certain that NBACC researchers do not violate the 1975 Convention, some scientists and government officials remain concerned. In the course of carrying out their mission, will NBACC scientists engage in the production of genetically modified pathogens themselves, in order to study them and find ways to defeat them? What about the accidental escape of these agents from the labs creating them? We know that accidental contaminations with the SARS virus caused several cases of this disease in Asia, and it is now clear that the 2007 outbreak of foot and mouth disease in Britain in 2007 was due to escape of that virus from a nearby research facility.

And could such pathogens ultimately find their way into the hands of disgruntled American scientists, as was likely the case with the anthrax spores used in Amerithrax? (Which, by the way, if true suggests that the United States is in fact engaged in their production, a violation of the 1972 Biological Weapons Control Act!) It has been argued that the very existence of such facilities further increases the likelihood of escape of potentially lethal agents into the environment.[8]

Other scientists think that the number of possible permutations of genetic modifications that could be made to the large number of pathogens already in existence, plus the creation of new ones, is essentially unlimited, and it will never be possible to anticipate all of them. But for now, at least, it seems the NBACC is going to try. What does all this mean for the ability of America to defend itself against bioterrorism in the twenty-first century and beyond? Will it be possible for terrorists to order up the components of the 1918 flu virus, assemble it, and unleash it on entire cities? Are we going to be facing viruses that escape immune detection in individuals previously vaccinated against the native form of the virus? What about inserting cassettes of genes conferring resistance to multiple antibiotics into the genomes of plague or anthrax bacteria? These are technologies that already exist, right here, right now.

A lot depends upon what federal regulatory agencies, and the scientists who use these technologies, do to make certain they do not fall into the wrong hands. A great many responsible scientists are concerned. Even private companies who make a lot of money generating the tiny RNA or DNA segments used to create viruses in the laboratory are eager to develop some sort of oversight mechanism, to ensure their services are not co-opted by would-be terrorists.

Some also would like to see controls placed on publication of data that could be used by terrorists. This may not be possible. The very kinds of technological advances that are cause for biosecurity concerns will underlie some of the most important medical advances we have ever known, and progress in this field depends on the free flow of information among scientists around the world through scientific and medical journals as well as internet databases. It is unthinkable to impede this exchange of information. These journals and databases are, however, open to the entire world, including scientists who may wish to harm the United States or its allies.

But even given the free availability of information about how to manipulate the DNA of potentially deadly pathogens for the wrong reasons, how likely is it that terrorists who would like to do us harm can actually use these technologies to create new superweapons to use against us? It is probably not as likely as Senator Frist once told the world:

A few technicians of middling skill using a few thousand dollars worth of readily available equipment in a small and apparently innocuous setting [could] mount a first-order biological attack. *It is even possible to synthesize virulent pathogens from scratch,* or to engineer and manufacture prions....[9] (Emphasis added.)

We'll return to this point in chapter 10.

CHAPTER 5

THE ULTIMATE BIOTERRORIST: MOTHER NATURE!

THE AGENTS OF BIOTERRORISM WE DISCUSSED IN CHAPTER 3 have a common feature that distinguishes them from other agents of terror, such as explosives, chemicals, and nuclear devices: they are themselves not the product of human invention. They arose in nature. Most of them, at one time or another, have given rise to events—often repeatedly, throughout history—that closely resemble what we imagine a bioterrorist attack would be like. These events occurred without human causation, as a result of the constant attempts of various microbes to invade the human body and use it as a place to rear their young.

In fact, the greatest threat we face from biological agents of death today is not from humans—it is from nature itself. A bioterrorist attack that could kill hundreds, possibly thousands, of Americans is a possibility, but, as we will discuss in chapter 9, one with low probability. A pandemic caused by a naturally occurring biological pathogen that could kill tens of thousands, possibly millions of Americans is an absolute certainty. Natural pandemics[1] are a regularly occurring phenomenon throughout history.

Historically, the most problematic pathogens for humans have been *Yersinia pestis* (plague) and the smallpox and influenza viruses. Our understanding of the natural ecology of *Y. pestis* in human and animal hosts makes it pretty unlikely we will ever

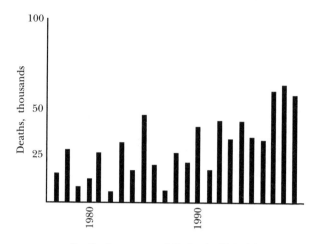

FIGURE 5.1 *Deaths from seasonal flu in the United States by year.*

again face a natural plague pandemic. The smallpox virus no longer exists in nature. But we remain, after decades of study, terribly vulnerable to killer flu pandemics. Even seasonal flu still kills on average between 35,000 and 40,000 people in the United States each year (Figure 5.1[2]). Three flu pandemics in the twentieth century—one major and two minor—claimed well over half a million lives in the United States alone, and almost certainly over 100,000,000 worldwide.

The uncertainty about exactly when a lethal flu pandemic could strike and the resulting social and economic disruption, in addition to the large numbers of deaths and illnesses even a moderate pandemic would cause, are in every way comparable to the uncertainty and consequences of a successful bioterrorist attack. Yet a reasonable wager would be that nine out of ten Americans fear the latter much more than the former. Why that might be will be a thread running through much of the rest of this book.

Our societal responses to a bioterrorist attack with the pathogens on the CDC lists are for all practical purposes the same as our responses would be should any of these agents, or any other pathogenic microbes, known or unknown to us now, make their way into the human population and create a natural epidemic or pandemic. The major difference between how such an event would affect us today and the situation even a hundred years ago is the presence in many countries of strong public health systems. These same public health systems will also provide our major defense against the results of a bioterrorist attack.

To give some impression of what a natural pandemic might look like, let's take a look at several situations of relatively recent history: the three influenza pandemics of the twentieth century, the SARS pandemic of 2003, and the still uncertain health crisis that could be caused by the H5N1 variant of the avian flu virus.

THE 1918 INFLUENZA PANDEMIC

We don't know exactly where the strain of influenza virus causing this pandemic arose. Most likely it was somewhere in Asia, probably China. It may have emerged as early as January of that year, or even earlier. Influenza outbreaks came in several waves during 1918. The first washed ashore in February in Spain, thus giving rise to the popular term "Spanish flu." This flu proved to be highly contagious, but the resulting disease was relatively mild. The one that would kill so many people didn't arrive until September.

Although the 1918 pandemic coincided with the latter months of World War I, emergence of the virus was not caused by the war itself in the sense that it was developed or used as a weapon. Certainly crowded, dirty trenches and a general lack of hygiene everywhere—especially military camps—hastened the spread of natural outbreaks in areas where the war still raged. Half of American fighting men in this war died of the flu—not from artillery shells, bullets, or poison gas. Mysteriously, the pandemic ended very shortly after the war.

When it was over, barely a year after it began, close to a third of the world's population may have been infected. We don't really know how many people died worldwide. Estimates have ranged from twenty-five million to fifty or even a hundred million—possibly as much as several percent or more of the world's population at that time. Precise medical record keeping was still less than perfect in the United States and Europe, and practically nonexistent in many parts of the world. An estimated twenty million people in the United States were infected, and between six and seven hundred thousand people perished—slightly more than have died after twenty-five years of AIDS, and more than have died in all U.S. wars through Vietnam. All evidence suggests that two or three pandemic-level flu events per century is the historical norm. The death rate in even the more serious seasonal flu outbreaks—the kind we see almost every year—is around a tenth of one percent of those infected; the death rate during the 1918 pandemic was twenty-five times that. Virulence at this level was not recorded before 1918 (although it may well have occurred) and has not been seen since. One of the striking features of this influenza outbreak is that so many of its victims were between twenty and forty years of age, in the prime of life (Figure 5.2); the flu is normally most deadly for the

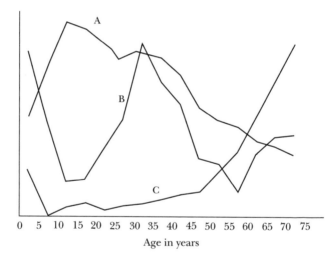

FIGURE 5.2 *Proportional distributions by age. (A) Rate of infections in the 1918 pandemic; the disease struck across all age groups, although the young were particularly susceptible. (B) Rate of deaths in the 1918 pandemic; surprisingly, the death rate in individuals 1–40 years of age was extremely high, by comparison with (C), the rate of deaths seen in a standard outbreak of influenza. (Based on data presented in Jeffery K. Taubenberger and David M. Morens, "1918 Influenza: The Mother of All Pandemics," Emerging Infectious Diseases 12(2006):15–22.)*

very young and the very old. The 1918 flu was also active during the spring and summer, whereas flu is usually a problem only in winter months.

The disease symptoms were like those of the flu generally— fever, achiness in joints and muscles, dizziness and weakness— but they were unusually harsh. Most people in fact recovered, but for too many others, death could come within just a few days of the onset of symptoms. Hemorrhaging in the lungs was common, causing victims to spit up quantities of blood-laced froth. As breathing became more difficult, many patients turned blue from lack of oxygen. Pneumonia could set in after a few days, was essentially untreatable, and was the most common cause of death.

All influenza viruses that have been involved in human epidemics or pandemics are of the influenza A viral group (Box 5.1). Influenza B and C viruses cause relatively mild cold- or flu-like symptoms in humans. Influenza A viruses are thought to have originated in aquatic fowl, and find a natural reservoir in many

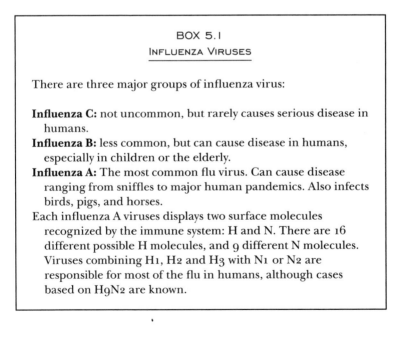

BOX 5.1

INFLUENZA VIRUSES

There are three major groups of influenza virus:

Influenza C: not uncommon, but rarely causes serious disease in humans.
Influenza B: less common, but can cause disease in humans, especially in children or the elderly.
Influenza A: The most common flu virus. Can cause disease ranging from sniffles to major human pandemics. Also infects birds, pigs, and horses.
Each influenza A viruses displays two surface molecules recognized by the immune system: H and N. There are 16 different possible H molecules, and 9 different N molecules. Viruses combining H1, H2 and H3 with N1 or N2 are responsible for most of the flu in humans, although cases based on H9N2 are known.

types of birds, where they live in the gut, causing little or no damage.

Like HIV, flu viruses use RNA rather than DNA to build their genetic blueprint. Unlike DNA viruses, RNA viruses do not "proofread" their blueprint when it is copied from one generation to the next, resulting in a very high mutation rate—a characteristic of both HIV and flu viruses. That is what has made it difficult to develop effective vaccines for both of these pathogens.

Current strains of avian flu viruses have come into balance with their bird hosts. They do not kill them, and can enjoy a long and productive lifetime within their hosts. Thus, most mutations of their genetic blueprint are harmful, disrupting the balance between virus and host, and are selected against. Nonetheless, occasional mutations do arise in avian flu viruses that result in highly pathogenic forms that result in destruction of large numbers of birds before the viral mutants eventually disappear.

Influenza A also infects pigs as well as humans. Humans are not ordinarily infected directly by A-type viruses that pass back and forth among birds, but occasional mutants arising in birds or in pigs can produce a variant that is able to infect human cells. In both cases, the virus settles into the lungs, where, being completely foreign, it is vigorously attacked by the immune system. The rapid mutation rate of RNA viruses is now an advantage, as it changes rapidly in a thousand different directions, trying to escape immune destruction and live at peace with its new host.

Within the A group, viruses can be further classified based on two types of molecules they all display on their surface: the H (hemagglutinin) molecule, and the N (neuraminidase) molecule. These are viral proteins involved in the initial infection of target cells (H) and the escape of newly formed viruses from infected cells (N). The H proteins interact with receptors on host cells and promote entry of the viruses into them. That is one reason true avian viruses rarely cross into humans: they are selected to recognize bird cells, not human cells.

Influenza A viruses have sixteen possible different H molecules they can display, and nine different N molecules. Each strain of influenza A is named based on the combination of these two molecules that it uses: H_1N_1, H_5N_1, H_7N_7, etc. H_1, H_2, and H_3 are most commonly found in viruses that infect humans, usually in combination with N1 and N2. Analysis of the recovered samples of the 1918 virus discussed in the last chapter showed it was of the H1N1 strain, and it is thought to have migrated directly from birds to humans.

But at the time of the 1918 pandemic, scientists hadn't even identified viruses as pathogenic agents distinct from bacteria—that would not happen until the 1930s.[3] The lungs of many who succumbed to the 1918 flu were full of bacteria, so it was assumed at the time this was yet another bacterial disease. We now know that heavy viral infections, particularly of the airways, are often accompanied by secondary bacterial infections, including the pneumococcal infections frequently associated with the pneumonia seen in late-stage flu. At any rate, since neither drugs that could slow the internal spread of the flu virus nor antibiotics to treat bacteria were available in 1918, it would have mattered little to have correctly diagnosed the causative agent. And of course there was no influenza vaccine available, or ventilators to assist with breathing.

The lack of any effective treatment also meant that public health measures were limited more or less to those that were present already in medieval times. Public health as a recognized medical discipline was still in its infancy in 1918, at least in the mind of the public, and compliance with official public health recommendations was irregular at best. It was understood that the disease was spread through aerosols generated by sneezing and coughing, and affected individuals were isolated at home or in segregated areas of hospitals. Entire groups of people and physical locations were often quarantined—military bases, prisons, and asylums, for example.

Hospitals overflowed with victims, and the scarcity of doctors and nurses caused by their recruitment into the war was compounded by the unwillingness of many who might have helped minister to the sick to even enter the hospitals. Medical and nursing students with little experience were often thrust into the fore of battles against the pandemic. Patients were confined to their beds, usually surrounded by bedsheets strung up to limit the spread of contagious aerosols. Both health care workers and patients wore masks. Isolation and quarantining of victims led to hostile interactions between the public and health and public safety officers. Disposal of infected bodies became problematic—some communities even resorted to open trench mass burials, which again met with strong public resistance.

In the hardest-hit communities, such as Philadelphia, there was severe social and economic disruption. In that city, 11,000 people died in the first month alone. The lack of any national uniformity of public health standards meant that each state, city, and county basically made up its own rules. Schools and businesses were closed in many cities, as were bars and theaters of all sorts. Public gatherings were banned. Churches generally remained open, but congregants were urged to sit as far apart as possible. In some cities, even large weddings and funerals were forbidden, and people were not permitted in government buildings or on public transportation without a face mask. It was a time of enormous chaos.[4]

Although the story of the 1918 influenza pandemic lacks the stridency and militaristic drama of exercises like Dark Winter and Atlantic Storm, it certainly stands alongside them in terms of damage done in terms of human lives lost, social collapse, and economic disaster. But one very big difference leaps out in this comparison: the 1918 pandemic actually happened. And tens of millions of people died.

THE PANDEMICS OF 1957 AND 1968

After the 1918 pandemic ended, flu reverted to its usual pattern of seasonal appearances. Most authorities think the pandemic H1N1 strain passed from humans into pigs, where it clearly was not as virulent. Milder versions of H1N1 must have arisen in pigs and passed back to humans, or mutated within humans themselves; at any rate, relatively benign H1N1 became the dominant FluA variant in humans for the next forty years.

Of course no flu variant or flu season is ever trivial; even today, with both vaccines and drugs to control the virus and antibiotics to manage secondary bacterial infections, tens of thousands of people in the United States (probably a million worldwide) still die each year from complications of seasonal flu. But the next flu outbreak after 1918 to rise to the status of a worldwide pandemic came in 1957, triggered by a variant of the flu virus originating in Guizhou province in China, where it first infected humans in 1956. It was commonly referred to as the "Asian flu."

Like most flu pandemics, including the one in 1918, this one was triggered by a viral variant that few living at the time had ever encountered. The prevalent H1N1 strain, probably while in pigs, appears to have interacted with an avian flu virus from ducks to produce an H2N2 variant. (The H and N genes and one other viral gene were from the bird virus; the rest of the genes were from the human virus.) This may be the first time a flu virus variant containing N2 had entered the human population in at least two generations, since there was no immunity to it among humans.

This new form of the virus was able to cause pneumonia in humans entirely on its own, without help from secondary bacterial infections of the lungs, although such infections did occur. Children were especially vulnerable to infection, and schools became major venues for spreading the disease. Many schools closed at least briefly, although relatively few children died. On the other hand, this virus was particularly deadly for persons with lung or heart disease, and took a high toll among the elderly. It was also very dangerous for women in the third trimester of pregnancy.

It has been suggested that this virus may have infected as many or more people as the 1918 variant, but the availability of a vaccine by May of that year,[5] as well as antibiotics and improved public health services, contributed greatly to a reduction in overall mortality. Still, the toll it exacted—70,000 deaths in the United States and 2,000,000 to 4,000,000 worldwide—was certainly horrendous.

The H2N2 Asian flu virus largely displaced the H1N1 variant circulating among humans after 1918 and continued to cause flu outbreaks for several years after 1957, with a particularly serious eruption in 1958 involving disproportionately the elderly. But as increasing numbers of people built up immunity to it, through immunization or natural exposure, it gradually became less of a problem. It essentially disappeared after 1968, when it was largely supplanted by the infamous Hong Kong flu variant.

The 1957 version of the H2N2 virus almost made it back into the human population in 2004–05. A U.S. laboratory accidentally included it in influenza test kits sent to other laboratories throughout the world. Fortunately, after an alert laboratory in Canada recognized the mistake, the kits were immediately recalled, and no one appears to have been infected.

Eleven years after the Asian flu, a third pandemic emerged from the Chinese mainland, becoming known as the Hong Kong flu.[6] This time the culprit was an H3N2 variant of the influenza A virus. (The H3 and one other gene were of bird origin; the rest of the genes were human.) It first came ashore on the West Coast in the United States, probably with troops returning from Vietnam. From there it spread eastward, although not all states were ultimately affected except for sporadic cases. Although about 20 percent of people in affected countries became infected (an estimated 50,000,000 in the United States), this was a relatively mild pandemic, with only about 34,000 excess deaths[7] in the United States and 700,000 to 1,000,000 worldwide. But once again, the elderly were hardest hit. Drugs that we have now to lessen the impact of flu—amantadine, rimantadine, ostelamivir (Tamiflu), and zanamivir (Relenza)—were not yet available. Less deadly forms of the H3N2 Hong Kong flu virus variant evolved over the next few years and are still the dominant form causing seasonal flu in humans to this day.

There are a number of reasons for the relative mildness of the Asian and Hong Kong flu pandemics. Neither variant in these latter outbreaks appear to have induced the same kind of violent response by the immune system, with collateral damage to normal tissues, thought to have been responsible for much of the damage seen with the 1918 H1N1 virus. Moreover, there had been continued improvements in flu vaccines used to control spread of flu viruses and in the antibiotics used to treat bacterial complications. The H3N2 variant of 1968 was immunologically cross-reactive with the H2N2 variant from 1957, and so people exposed to H2N2 over the intervening years may have had some degree of immunity to the Hong Kong virus. Also, the Hong Kong virus struck many American cities right at the December-January school break, reducing transmission among students. Some schools and colleges closed slightly early or delayed re-entry until the flu had subsided to reduce spreading to the larger population.

So the twentieth century saw early on one of the deadliest flu pandemics in recorded history, and later gave rise to two more pandemics of decreasing intensity. Does this mean that flu pandemics are on their way out as a threat to humans? Would that it were so! We are now staring down the barrel at a new flu variant coming to us from birds that has the potential to equal or exceed the devastation wrought by the 1918 virus. We first saw this virus—this time an H5N1 variant—in 1997. But between that time and the present, we went through another pandemic with a completely different virus—SARS. Before we look more closely at H5N1 flu, let's take a moment to explore what happened in the SARS pandemic of 2003–04.

THE SARS (SEVERE ACUTE RESPIRATORY SYNDROME) PANDEMIC OF 2003–04

The first cases of what would come to be known as SARS were detected in China at the end of 2002. Several people in Guangdong province showed up at clinics and hospitals with severe flu-like symptoms. At the time, there was nothing to mark this as the emergence of a new and deadly disease that would eventually affect over 8,000 people in thirty countries, killing 774 of them. These numbers, from official World Health Organization records, may not tell the whole story. It is likely that more people in China were involved, and may have died, but either went unnoticed and unrecorded or were incorrectly diagnosed. There is evidence that SARS was present in China prior to 2002. For example, a subsequent analysis of 938 blood samples collected in Hong Kong in 2001 for unrelated purposes showed evidence that 17 of these individuals had been exposed to SARS.

The first few patients subsequently identified as having SARS apparently recovered, and may not have passed the disease on to others. But within a month or two many other such cases showed up in China and began to worry authorities. It takes time to figure out, in a situation like this, that what one is seeing in the clinic is not just another outbreak of a slightly nastier flu, but a new, different, and more deadly disease. China's public health system, while adequate, is not yet quite on a par with most Western countries. That fact, coupled with the penchant of Chinese authorities for trying to keep disturbing news out of the press,[8] slowed the development of awareness in the rest of the world that a major new health threat was arising. China did not alert the WHO until February 2003, by which time several dozen cases had been detected. China subsequently issued a public apology for its slowness in dealing with the SARS crisis.

Hong Kong was particularly hard hit early in the developing pandemic, as was Singapore. SARS had minimal impact in the United States. Only eight people were diagnosed with SARS; all of these had recently been in countries with verified SARS outbreaks. None died. However an American businessman traveling in Asia in early 2003 died of SARS in a Hanoi hospital. There was, moreover, a more serious importation of SARS into Canada, particularly Toronto. Of 251 cases officially diagnosed, 44 died. Nearly half of these were hospital or other medical personnel, and some hospitals in that city had to be quarantined. A high percentage of the 300 deaths in Hong Kong were also among health care workers. The public health systems in both cities were all but brought to their knees.[9] Both cities also experienced serious economic disruptions and restrictions on travel.

These events, together with a better picture of what was happening in China, contributed to the sounding of a global health alert by the WHO. Public health agencies in countries throughout the world then issued alerts of their own, and began the serious work of preventing further spread of this new disease. The SARS pandemic peaked in May 2003, rapidly subsiding as the resulting containment measures took effect. The epidemic was essentially over by mid-2004 after a brief recurrence in China.

SARS is spread either through the air by sneezing and coughing aerosols or by direct contact with bodily fluids. As with many other contagious respiratory diseases, symptoms usually appear within two to three days of exposure to the triggering pathogen. The causative agent in the case of SARS is a newly emerged variant of the coronavirus, one of many viruses that can induce human colds. This never before seen variant, now called SARS-CoV, was first identified in March 2003 as a possible cause of SARS. An extraordinarily intense WHO-financed investigative campaign in laboratories around the world quickly confirmed this, and on April 16, a month before the pandemic peaked, the WHO announced SARS-CoV as the official causative agent in SARS.

SARS-CoV was found a short time later in bats and civets in Guangdong province. It seems likely that SARS-CoV jumped to humans either through bat bites or through eating the meat of civets or other small mammals. Meat from civets, a carnivorous cat-like animal common in Asia, can be found in numerous meat shops throughout Guangdong. Whether bats or civets are a natural reservoir for SARS-CoV is unclear.

The symptoms of SARS are similar to the flu: fever, generalized achiness, lethargy, abdominal discomfort. There is usually a dry cough early on, and there may be shortness of breath, both of which would be unusual for the flu. Fever usually peaks in most patients four days after onset of symptoms, lung abnormalities are revealed by X-ray at day six, and oxygen sufficiency may be critically low at day 8. This latter may be particularly crucial in people over sixty, where fatality rates often approach 50 percent. In children and young adults, fatalities rarely exceeded 10 percent of those infected.

There was no vaccine available for this previously unknown viral variant, and flu drugs were ineffective against it. So the only interventions available at the time were traditional public health measures: making people aware of the symptoms and encouraging early self-reporting; identifying and isolating infected individuals and their first-degree contacts; urging the public to increase personal hygiene, and to wear face masks where appropriate; encouraging "social distancing" (avoiding mass gatherings) and closing schools where necessary; increasing surveillance at ports of entry for symptomatic individuals. Travel advisories warning people away from infected areas such as Toronto and Hong Kong were also issued by most governments.

Some of these measures seem to have been effective in places like Hong Kong, Singapore, and China proper. Government authorities in these cities tend to be a bit more heavy-handed in enforcing public health edicts, and this may have been effective in limiting the contagion emerging from these areas. It appears to have been less effective in Toronto. Only a few of the milder recommendations were issued by U.S. officials. An analysis of the effectiveness of these various measures in containing the SARS pandemic has been published.[10]

SARS is a much slower moving infection within a population than the flu, offering more opportunities for intervention. But there is in fact no rapid, clear test that could be used in a doctor's office or small clinic to determine whether someone with what appears to be a bad case of the flu may actually have SARS. As the 2002–03 pandemic progressed, the major factor in determining whether people presenting themselves at a clinic or doctor's office with serious flu-like symptoms might have SARS was whether they had recently been in contact with someone who did have SARS or had traveled in a location where SARS was prevalent.

Laboratory testing could now conclusively identify SARS-CoV as the causative agent, but anyone even suspected of having SARS would likely be immediately placed in isolation, closely monitored, and given intense supportive therapy as needed until testing was completed. SARS cases can proceed rapidly into severe breathing difficulties and an insufficiency of oxygen, leading to respiratory collapse (acute respiratory distress) and a requirement for ventilator-assisted breathing.

As with most diseases caused by viruses, there is at present no specific treatment to control SARS-CoV. Indomethacin and interferon have been reported to be effective in the laboratory, but there is as yet no supporting evidence for the efficacy of these drugs in the clinic. There is also no vaccine presently available that protects against SARS-CoV, although several promising vaccines are under development, and one is currently being tested in human clinical trials being supervised by the U.S. National Institutes of Health. Researchers throughout the world are working steadily to develop a defensive armamentarium should the SARS virus ever reappear.

OUR WORST NIGHTMARE? THE POSSIBILITY OF AN H5N1 FLU PANDEMIC

As noted earlier, influenza A viruses live most of the time in the gut of birds without sickening or killing them. But occasionally, random mutations arise in these viruses that can trigger a severe form of avian flu, killing large numbers of birds. In the wild, these mutations can wreak havoc in isolated flocks, but the resulting explosions are usually self-extinguishing given the large spacing between flocks and their constant movement from place to place. Once the infected birds or even whole flocks die out, the lethal variant of the virus can disappear.

But modern methods of rearing poultry for commercial purposes result in huge concentrations of birds in very constricted spaces. Spread of mutant viruses can be extremely rapid in such populations, and the only solution to containing an outbreak is the immediate, compulsory destruction of every bird in the infected compound, whether symptomatic or not. Even then, shipment of infected live birds to other locations prior to confirmation of an outbreak (or even afterwards, in the interest of limiting financial loss), as well as movement of contaminated equipment, truckers and other workers between farms, can result in entire regions having to destroy enormous numbers of birds.

Such was the case in Hong Kong in 1997, with the emergence of a deadly avian flu variant of the type H5N1. This variant did not sit benignly in the gut of birds it infected, but penetrated into every organ and tissue of the body. This is thought to be due to changes in the H molecule, which determines which cell types the virus can invade. The result was rapid physiological collapse and death. In former times such outbreaks were referred to as "fowl plague." Only in the 1950s were they recognized as a form of avian flu. Since that time there have been a dozen or so outbreaks, usually involving viruses bearing the H5 or H7 forms of hemagglutinin.

The 1997 Hong Kong outbreak, which involved several poultry farms, was finally quashed after the destruction of tens of thousands of birds, and it seemed that H5N1 would likely fade into the sorry history of fowl plague. But in May of that year, a three-year-old boy in Hong Kong was admitted to a hospital with a respiratory infection that quickly progressed into pneumonia. But as with birds, the H5N1 that had infected him spread far beyond the lungs. He went on to develop Reye's syndrome, acute respiratory distress, and kidney and liver failure. He died a few days later. The medical staff at the hospital, stunned by the violence of his disease, were determined to find out what had caused it. Throat washings taken from the boy had been saved, and were analyzed for a wide range of viruses and bacteria. None of the tests detected anything. Samples were sent to WHO labs in London and Rotterdam and to the CDC in Atlanta. The Rotterdam lab was the first to come back with an answer: the throat washings were positive for the H5N1 variant of the avian influenza virus.

Since H5N1 had never been seen in humans before, extensive tests were carried out to see if a mistake had been made or the throat washings had been secondarily contaminated. Several labs around the world joined this effort, and it was soon absolutely clear that the young boy had indeed succumbed to a primary infection with the H5N1 virus. Further analysis of the boy's virus showed that it was virtually identical to the H5N1 flu virus involved in the recent local poultry outbreak, and that it had passed to him without modification in an intermediate host such as pigs. The clinical description of his illness and death were hauntingly familiar to those who had studied the 1918 flu pandemic.

Public health authorities in Hong Kong immediately tested other members of the boy's family, as well as medical staff that had attended him in the hospital. None of his family members showed any signs of having been in contact with the virus,[11] but one of his nurses and some of his playmates did. Wider testing picked up a number of poultry workers who also showed signs of having harbored the virus. None of these had showed any signs of a flu infection. But when samples of the virus isolated from the boy's throat washings were tested on an experimental poultry flock in Georgia, the entire flock underwent an immediate and violent death from catastrophic influenza. From that point on, it was agreed that all work with the Hong Kong H5N1 virus must be carried out in high-security biocontainment laboratories, such as those designed to work with soil samples brought back from the moon or with CDC A-list pathogens.

In the months after the boy's death, no further cases emerged, and public health authorities began to hope his death might have been a fluke. It was unclear how or even whether he had passed the virus directly to others around him. It could not be ruled out that those who had contact with him and showed positive for

H5N1 had not independently picked it up from poultry. And he was the only one to have become ill.

But the message to infectious disease specialists was clear. Here was a virus that could—did—kill a human being, and it had passed directly from birds into humans. What would happen if this new H5N1 virus infected a human during the normal flu season, and that person was also carrying one of the relatively benign seasonal flu viruses such as the current H3N2 variant? The viruses could well recombine, producing a hybrid variant with H5N1's lethality and the ready human-to-human transmissibility of a seasonal flu virus.

A few months later, any hope that the death of the three-year-old child had been a one-shot occurrence disappeared forever. Over a period of several weeks, a total of seventeen more children and adults were admitted to hospitals with signs of a violent pneumonia and were found to be infected with H5N1. Five died. As in the 1918 pandemic, it appeared that most of those who died were previously healthy adults. That was the only sense in which the initial three-year-old victim may have been a fluke. Most of those who became seriously ill in this round had had contact with poultry; there was still no reason to suspect human-human transmission of H5N1.

And then, almost immediately afterward, poultry began dying again on Hong Kong farms, and even in live-chicken markets in the middle of the city. Public health authorities moved swiftly. As soon as H5N1 was confirmed as the cause and it became apparent that up to a quarter of the territory's poultry were infected, they decreed that all of Hong Kong's poultry must be destroyed—over a million and a half birds. The economic cost would be staggering, but there was no hesitation. The government ordered it done immediately.

Since its initial detection in Hong Kong in 1997, the H5N1 variant has spread to birds in over fifty countries (Figure 5.3). As of April 2007, the WHO had confirmed 291 cases of transmission to humans, with 172 deaths. A partial history of the spread of H5N1 and its interaction with humans is shown in Table 5.1. The mortality rate in humans known to be H5N1-infected has hovered at about 60 percent for the past decade, making it at least twenty-five times more deadly than the 1918 H1N1 flu virus.

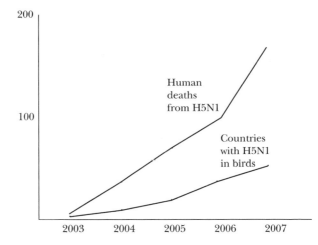

FIGURE 5.3 *Cumulative spread and human deaths from H5N1 avian flu.*

Victims so far have been predominantly women and children, but this could simply reflect their representation among individuals handling poultry.

The first (and so far only) confirmed case of human-to-human transmission occurred in Indonesia in 2006. A child contracted H5N1 from handling poultry and subsequently passed the virus to six other family members. No further spread outside the family was documented.[12] It may be that this family had a genetic alteration that made them more susceptible to human-human transmission. But it is this possibility—a genetic alteration in the virus itself, particularly in the H protein which determines initial entrance of the virus into cells—that gives world public health officials nightmares. It is what happened in 1918, and it could happen again.[13]

How would we handle an outbreak of transmissible H5N1 influenza in the human population? The ideal solution would be a vaccine that is at least as effective as the current vaccines for seasonal flu. We don't have such a vaccine yet, but one that may offer partial protection has recently received FDA approval. The vaccine used an inactivated H5N1 virus isolated from humans infected through contact with birds. In healthy individuals, antibody levels judged sufficient for protection were induced in about

TABLE 5.1 A Partial Summary of H5N1 Avian Flu Outbreaks Affecting
Humans

When	Where	What
1997	Hong Kong	Epizootic in poultry. Eighteen humans infected, probably from poultry. Six died.
2003	China, Hong Kong	Two Hong Kong family members returning from China developed H5N1 pneumonia. One died. Method of infection unknown.
2003–04	Thailand, Vietnam	Epizootic in poultry; first case of possible human-human transmission reported.
2005	China, Cambodia, Thailand, Vietnam, Indonesia	Epizootics in poultry; first instance of possible infection of humans from ducks.
2006	Azerbaijan, China, Cambodia, Egypt, Djibouti, Thailand, Iraq, Turkey	Epizootics in poultry; swans may have been source in some cases.
2006	Indonesia	First confirmed instance of human-human transmission. A child who probably picked up the virus from poultry passes it on to six other family members.
2007	India	Two separate poultry farms infected.
2007	Africa	Togo becomes seventh African country with H5N1 outbreak.
2007	Great Britain	Over 100,00 H5N1-infected birds culled

*For a complete timeline of avian influenza, see the WHO website http://www.who.int/csr/
disease/avian_influenza/timeline_2007_04_20.pdf.*

*As of April, 2007, 291 confirmed cases, 172 deaths (see http://www.who.int/csr/disease/
avian_influenza/ country/cases_table_2007_04_11/en/index.html for updates).*

half of those immunized. Current seasonal flu vaccines protect 90 percent or more of recipients.

And to achieve even this level of protection required two injections of a rather large amount of the vaccine, spaced one month apart. So this vaccine is far from ideal, but it is all we have at the moment, and it is being stockpiled as a stopgap measure by the federal government while research into a more effective vaccine proceeds at a rapid pace. Enough of the recently approved vaccine for twenty million people is currently available. University and drug company researchers are working furiously, supported by federal funds, to develop a more effective vaccine. The government has already drawn up a prioritization list for who will receive H5N1 vaccines during an epidemic crisis.[14]

There are four drugs currently approved by the FDA for mitigating the impact of influenza A infections (Box 5.2). Amantadine and rimantadine stop the virus from replicating after it enters the cell; Relenza and Tamiflu block the ability of newly made viruses to exit from the infected cell. There is already evidence that amantidine and rimantidine are ineffective against the H5N1 virus as it presently exists. Relenza and Tamiflu are effective against current H5N1 strains, and are being stockpiled by both the federal government and individual states.

But the problem for both vaccines and antiviral drugs is that the H5N1 variant that presently exists, even in those cases where it has migrated into humans, is not the one we have to worry

---

BOX 5.2
FDA-APPROVED ANTI-INFLUENZA DRUGS

Generic name	Trade name	Approved
Amantadine*	Symmetrel	1966
Rimantadine**	Flumadine	1993
Zanamivir	Relenza	1999
Ostelamivir	Tamiflu	1999

*Available as generic
**Available only as an inhalant

about. As a minimum, some sort of mutation will have to occur in one or more genes of existing forms of H5N1 to allow it to spread easily among humans. A second mutation, allowing it to jump more readily from birds to humans, would really put us in deep trouble. It is possible, some think even likely, that these mutations could alter the ability of current H5N1 vaccines to block the virus. The effect of these mutations on the sensitivity of the new mutant(s) to antiviral drugs will also have to be assessed.

So we won't really know what we are dealing with, what it is we will have to protect ourselves against, until such mutations actually occur. The new mutants will have to be isolated and studied in the laboratory, and plans devised for the most effective vaccines and drugs. Laboratories and manufacturers around the world are poised to do this, and to do it on a 24/7 basis. But it will still take time. It is entirely possible that when such a mutant arises, we will be on our own for as much as six months before drugs or vaccines are available.

How bad could an H5N1 pandemic be? We won't know that either until we have the miscreant mutant in hand. It could be more lethal than current H5N1 strains, or less lethal. If it maintained its present lethality, a third of the people in the world became infected, and half of those died—the math is pretty straightforward. Remember, H5N1 as it exists now, when it infects humans, is about twenty times more lethal than the 1918 virus. Dr. Anthony Fauci, Director of the National Institute of Allergy and Infectious Diseases, considers the threat of an H5N1 flu pandemic greater than that of bioterrorism.[15]

We have the experience of those who managed the 1918 outbreak with nonpharmaceutical interventions to guide us.[16] And it is possible that with our improved ability to manage influenza-like diseases—we now have mechanical ventilators to assist with breathing[17]; we know better how to manage secondary respiratory bacterial infections and pneumonia; we will (hopefully) have anti-flu drugs—we could reduce the lethality of infections considerably. Maybe we could cut it in half. But that could still mean a billion people dead worldwide, fifty million in the United States. That would pretty much bring the world as we know it to a standstill.

There is reason to be concerned.

CHAPTER 6

AGROTERRORISM

The Very Food on Your Plate
(and the Water in Your Glass)

IN THE DARK WINTER SCENARIO WE REPLAYED IN CHAPTER 1, you may have overlooked one of the items placed on the hypothetical agenda that was disrupted by events in Oklahoma City and elsewhere: Taiwan had accused China of having purposely spread foot-and-mouth disease virus over several pig farms. Animal and crop pathogens have been a part of the biological warfare armamentarium of nearly every country that ever had a bioweapons program. Such an attack could certainly be plausible.

In the context of something like a smallpox attack, you might think this seems like small potatoes, barely getting onto the radar screen. You'd be wrong. If the true purpose of bioterrorism is social and economic disruption and the spread of fear and uncertainty across as many people as possible for as long as possible, disruption of food and water supplies can be a major weapon. That form of bioterrorism is called agroterrorism.

Agroterrorism is a threat not only to the daily bread and meat that goes on our plates and the water we wash it down with. Serious interruption of our domestic food or water supply is actually a rather low-probability outcome of an agroterrorism attack. The major impact would be economic. The United States is a major supplier of food to the world, and agricultural exports account for a sizeable portion of America's trade income—about $50,000,000,000 per year. Contamination of American food supplies, which feed a sizeable portion of the world community, could bring about economic disaster the likes of which have not been seen since the Great Depression. And it would affect not just farmers and agribusinesses who grow food for export. The collateral damage stemming from interference with this sector of our nation's business would ripple through the rest of our economy like a hot knife through butter.

ANIMAL TARGETS OF TERRORISM

To get a feel for what an agroterrorism attack could do, let's create our own Dark Winter scenario—we'll call it Dark Summer, since that's when an agroterrorism attack would most likely take place. And we'll replace smallpox released in shopping malls with one of the deadliest animal pathogens, foot-and-mouth disease virus, released on four major agribusiness cattle farms in different states.[1]

One hot July day in 2003 workers on one of these farms notice cattle starting to limp as they move around the fields. On another farm, cows that are being milked balk when hooked up to the milking machines, and workers notice blisters on their teats. Veterinarians are called, and discover blisters inside the mouths of the cattle as well. Like physicians with smallpox, these vets have never seen this disease before except in textbooks, but immediately recognize it for what it is: foot-and-mouth disease (FMD), last seen in the United States in 1929. The vets immediately alert the CDC; over the next thirty-six hours, veterinarians from all four sites have reported in. It is clear this is a terrorist attack, although whether foreign or domestic terrorists are involved is unknown.

The President is alerted by the Secretaries of Agriculture and Health and Human Services, and immediately calls his National Security Council into emergency session. He cancels all but his most essential business for the next two days. The President and his council are joined by CDC officials and the Secretary of Agriculture, who arranges for an expert on foot-and-mouth disease to address the assembled roomful of officials. Other key personnel are invited as well. The FMD expert rises to speak.

Foot-and-mouth disease, she tells them, is caused by a virus, called simply foot-and-mouth disease virus (FMDV), related to the polio virus that affects humans. It is more infectious than the smallpox virus. It infects cloven-hoofed animals: cattle, pigs, goats, and sheep among domesticated animals, as well as deer and elk in the wild. The disease is highly contagious but affects different animals differently. Among domestic herds, pigs and cattle are highly susceptible, sheep and goats less so. The latter two get only mildly sick at most, and can serve as a reservoir for the virus, which must be taken into account in any containment strategy. Humans are not made sick by FMDV, but they can transport the virus on their person for several days, and can spread it through inhalation and aerosol generation. This, too, must be taken into account.

Adult pigs and cattle do not usually die from FMD infections, but they become ill and severely disabled. They cannot reproduce, cannot nurture their young, and cannot be used as a food source, since all of their tissues and organs are laden with virus that could find its way back into the environment. The adults usually recover in two to three weeks; younger animals may die. The disease spreads rapidly, primarily as an exhaled aerosol, since the lungs, too, are full of virus.

Since FMD has been absent from the United States for nearly seventy-five years, American cattle, like most American citizens when it comes to smallpox or H5N1 influenza, have no natural immunity to FMD. They are also not routinely immunized against FMD. Given the absence of the virus in North America, it would be difficult to justify the expense of maintaining an active vaccination program, particularly since the currently available vaccine does not induce strong immunological memory. Periodic booster shots are required to maintain good immune defenses, which would add to the cost of such a program. We are not alone in this situation. Nearly all of Europe—at least the membership of the EU—has exactly the same policy.

"Another complication," she says, "is that there are about sixty different forms of the virus; each would require a different vaccine. Also, current FMD vaccines do not prevent infection, but protect the animals enough to keep them from becoming sick. That means vaccinated animals that are later infected, but seem to have resisted the virus, can continue to harbor the virus and spread it to other animals. Since they are not sick, it is impossible to spot them. Animals leaving farms or feed yards looking perfectly healthy could still be carrying the virus, and they can continue to carry it for a year or more. For that reason, no country that is FMD-free will accept animals or any parts of animals from vaccinated herds. So the United States does not vaccinate, and in fact we don't even make a vaccine in this country."

A hand goes up from the floor. "So if we don't have a vaccine, how do we stop this damned thing?"

"I didn't say we don't have a vaccine; only that we don't make it here. There is a collection of vaccines based on various forms of killed FMDV, made in other countries and stored at a vaccine bank on Plum Island, off the coast of Long Island in New York. It is possible that in some scenarios of containing an outbreak, we might want to resort to a vaccine."

Another participant chips in. "So I take it vaccination is not your first choice in the current situation. What is?"

"Standard procedure at the present time is to immediately quarantine all cloven-hoofed livestock on a farm or feedlot where FMD is suspected, and on all farms within a three-kilometer radius. No animals in or out. If laboratory tests confirm FMD, all cloven-hoofed livestock on those lots or farms will be destroyed. Any cloven-hoofed wildlife found in the vicinity of those lots or farms will also be destroyed. The resulting carcasses will be buried or burned. What we are doing is similar to the 'ring vaccination' procedure that might be used in an outbreak of smallpox or pandemic flu in humans. We are trying to create in essence a firebreak around a raging contagion.

"The farms themselves will have to undergo extensive chemical-based decontamination. The virus will be everywhere: in manure heaps, on equipment, in the soil. These farms will be closed down for many months. Also, no persons will be allowed to leave farms where destruction of livestock is ordered without undergoing a thorough decontamination procedure themselves. As I indicated earlier, humans can definitely carry the virus to other livestock areas. We may have to consider quarantine of such persons until they are tested, to be sure they are not carrying.

"We will expand this process in new rings, or enlarge older rings until we have each outbreak under control. The resources that may have to be committed could be huge. We may need help from the National Guard, for example. I would just remind all of you that we have about a hundred million head of cattle in this country, and around seventy million pigs.

"If there are no further questions, I'll pass you back to the Secretary, who will brief you on anticipated political and trade issues."

The Secretary of Agriculture spreads some papers in front of him, and rises to speak.

"The tests Dr. Hofman talked about have already been performed, and are positive at all four sites. Livestock on those farms are now being destroyed. While most of America will understand this has to be done, the images that we will all be seeing on television and in newspapers and magazines in the coming days are going to be very upsetting, and will likely generate some domestic discontent. We must make every effort to communicate to news media outlets the absolute necessity of these grave steps, and ask them to help us educate the public to this fact.

"These images will also be seen worldwide, of course, and in some quarters are likely to generate as much celebration as commiseration. They will be on CNN and BBC, but also Al Jazeera. What will be shown, of course, is not terrorists attacking and contaminating American food supplies, but Americans killing, burning, and filling mass graves with American livestock. It will be an ugly sight, and a story that can be told in several ways. It surely will be.

"The farms involved in the first-round action sites, two of which are actually feedlots, have an estimated 220,000 affected livestock. Most of these are on the feedlots; these are our weakest link in a situation like this, because of the large number of animals and extreme crowding. We're lucky more feedlots weren't hit.

"It will take several days to slaughter all of the livestock involved. We fully expect that several, perhaps all, of the infected sites will spill virus into surrounding farms. These will be intensely monitored, and any sign of disease or any positive tests in even asymptomatic animals will lead to imposition of a three-kilometer circle and destruction of all at-risk livestock. We simply cannot predict at this time where this process will stop. We can also expect considerable resistance to wholesale slaughter from affected farm families and from corporations that operate most of the larger pork and cattle facilities. All of them stand to lose a great deal, even though we have compensation programs to help them through this.

"The initial effect of this on the domestic food supply will probably be minimal. We have already begun tracing animals that have been shipped from the affected farms, and these will be pulled back and destroyed, as will all susceptible livestock or other animals with which they have come in contact. Any meat, milk, or other animal products processed from animals at these farms during the past two weeks will be recalled and destroyed. It is likely we will still see some leakage of virus from these animals and animal products into the general environment. The entire agricultural and veterinary apparatus in this country is now on high alert to spot additional outbreaks.

"But many people will very likely stop eating any and all meat from susceptible species, even though infected meat, even if eaten raw, would have no effect on humans, other than to possibly make them at least temporarily carriers of the virus.

"More devastating for us, of course, is that all foreign countries will—within a day or so, if they haven't already—immediately block all imports of American animal and other farm products. This will extend beyond the affected species, since poultry and other farm animals could be harboring virus. It will also extend at least several months after our last confirmed case. Most industrialized nations, like us before this happened, are and want to remain FMDV-free. And no FMDV-free country will accept FMDV-vaccinated livestock, either, by the way. We have no complaint; we would do exactly the same if this were happening in another country.

"We may also see some interference with U.S. citizens traveling to FMDV-free countries, unless they can prove an urban (as opposed to rural) residence here. It is possible they may be required to undergo some sort of decontamination procedure or even quarantine. We'll have to see how this plays out. And

FMDV-free countries will likely discourage their citizens from visiting the United States on anything less than urgent business, and subject them to various decontamination procedures upon their return.

"The economic cost of this attack, although largely invisible to the public, at least at first, has the potential of being huge. I would remind you that in the aggregate, agriculture is one-sixth of our total gross domestic product—well over one trillion dollars.

"We will be faced with not only the cost of a containment program, but the loss of a very large sector of our export market, plus the cost of importing animal products for our own markets if a sizeable portion of our own population also boycotts American meat. Transporters, distributors, restaurants—all are going to be hit hard. Layoffs will be inevitable. The final costs, of course, will depend on how far this contagion spreads and how long it takes to control it. We can anticipate that it will be at least some number of months after we have contained it before our agricultural exports are accepted in FMD-free countries. We will lose tourism dollars. The final cost will be well into the billions."

We'll leave the assembled officials to their labors. We don't really need to run this play to its final act. Various government exercises carried out to simulate an agroterrorist act with FMDV have generated some possible outcomes. Crimson Sky estimated that thirty-eight million animals could end up having to be slaughtered in a moderate attack. This is about the number of animals slaughtered yearly for domestic and export consumption. Silent Prairie concluded that even a limited FMD attack would likely spread to forty states before it could be fully contained, with a loss of nearly fifty million animals. So somewhere between thirty million and fifty million livestock could have to be killed. No one is quite sure how we would do that, or even if we could. But it would have to be done.

Neither of these exercises appears to have included the cost to the American economy due to shutdown of U.S. food exports, which would certainly be total. No country would agree to import anything that had been on an American farm. The loss of jobs in various industries tied to agriculture would be enormous. Many of the farms remaining in family hands would likely be repossessed or sold. It would take years to rebuild herds.

Great Britain went through a real-life Dark Summer in 2001, except the FMD outbreak was natural, rather than resulting from a terrorist attack. No one is entirely certain how FMDV got into the country, which had been FMDV-free since 1967. Most assume it was through illegal import of meat from a country with ongoing FMDV problems. Initially, the virus infected pigs on a single farm. Sheep present an adjacent farm then became infected, probably through a virus-laden aerosol, and were shipped to other agricultural locations before the outbreak had been confirmed.

A pandemic took off from there. British authorities used a ring containment procedure like that described above: quarantine, followed by slaughter of all animals in infected herds.[2] The procedure was fully publicized in all media outlets, and the constant pictures of vast funeral pyres of slaughtered animals were traumatic for a great many people. The outbreak was 95 percent contained in about three months, but not fully extinguished for nearly a year. Humans and vehicles were later found to have been major carriers, and contributed significantly to spread of the disease.

The virus spread to at least two other European countries (France and Holland), probably through infected animal products processed and shipped before the outbreak was recognized. Britain suffered a total boycott of its agricultural products for many months. Many of us returning by air from Britain remember being questioned by our own agricultural people about where we had visited, and being made to step through pans of disinfectant as we passed through Customs.

Before it was over, Britain had destroyed nearly eleven million animals on over nine thousand farms. The estimated cost to the national economy has been estimated at between forty and fifty billion dollars. Most of this loss was from collateral damage in the non-agricultural sector.

AGRICULTURAL CROPS AS
TARGETS FOR TERRORISM

Terrorist destruction of agricultural crops is at once simpler and more difficult than bioterrorist attacks on humans or animals. It is simpler, because America's croplands are enormous and open. The number of acres that are under the watchful eyes of a farmer and his family, who might detect the presence of a stealthy stranger scattering strange powders or loosing insects (either alone, which would be damaging enough, or carrying a plant pathogen), has become vanishingly small. Soybeans, for example, are planted in over seventy million acres in the United States. Many of the largest agricultural fields are unguarded and largely unobserved most of the year.

Yet their very vastness makes it difficult to infect more than a small portion of the nation's crop at any one time. A crop dusting airplane might extend a terrorist's reach, but would raise immediate suspicions on even the largest, most remote farmlands. And while there are pathogens for plants as deadly and inexorable as any for humans, they are harder to detect, have longer incubation times before disease appears, and tend to spread more slowly. They can take months or even years to be detected and verified. Watching plants die, like watching them grow, doesn't make for very exciting TV coverage.

But while the psychological effect of an agroterrorist attack on plant crops might not reach the level of seeing animals slaughtered and burned every night on the news, the damage ultimately done to our economy and to our social structure could be nearly as significant. The television coverage would be less dramatic for the terrorists as well, of course, but depending on their aims—they could be domestic as well as foreign, after all—destruction of major portions of American agricultural crops could certainly have the desired effect. Although the possibility of the domestic food supply being seriously disrupted, at least during the year of an agroterrorist attack, is slim, there could be substantial follow-up disruption. The costs of destroying tens of thousands of acres of crops, and ensuring eradication of the pathogen, could be huge. Export markets could be devastated. Layoffs within the agricultural industry itself, economic disruptions in a major sector of the economy rippling outward, decreased confidence in the food supply, fear of eating contaminated food,[3] all are possible consequences of a terrorist attack on our crops.

It is also possible that terrorists could attempt to contaminate our crop food supply not in the fields, but at various points in the harvesting, processing, and distribution systems. This has the advantage of having large portions of our crops or their products concentrated in one place, the crop equivalent of a feedlot, in a sense, rather than scattered around the nation. And it is not entirely necessary that terrorists even be in the country to spread crop pathogens; they could be inseminated into fertilizers or other crop-enhancing products imported into the United States.

The pathogens terrorists might use to destroy crops are largely those that destroy crops naturally each season in the course of a normal agricultural cycle. Existing, naturally occurring plant diseases already cost the U.S. economy over $30,000,000,000 each year. *Tilletia indica* causes a disease of wheat called Karnal bunt. A major (natural) outbreak of this disease in 1996 threatened the entire U.S. wheat export market. The infestation was brought under control, at great expense and with considerable crop loss. But all of this is factored into the cost of doing business; we already pay for it at the supermarket. On the other hand, orchestrated, nationwide terrorist attacks with *T. indica* could well bring the domestic and export wheat markets to their knees in this country, with an incalculable economic impact that would affect every American.

AND ABOUT THAT GLASS OF WATER . . .

Not only did medieval armies lob rotting carcasses over castle walls, they also stuffed them into the wells and cisterns of their enemies. People can go weeks without food before succumbing, but only a few days without water. Public health authorities, as well as terrorism experts, have warned repeatedly that the nation's water supply is a potential target for bioterrorists. The potential for serious damage is clearly there—to human health, to agriculture, and to every sector of our economy that uses water. And in the end, that is virtually every sector.

There are over 160,000 water systems delivering water to U.S. customers; 350 of these serve populations of 100,000 or more, and are considered the most likely potential targets for terrorism. About half of our water comes from underground sources, and half from surface sources. The federal agency charged with making sure rotting corpses or other possible bioterrorist weapons are not stuffed into our drinking water supply is the Environmental Protection Agency (EPA). Title IV of the 2002 Bioterrorism Act directs the EPA, in coordination with the Department of Homeland Security, to work with all water utilities serving more than 3,300 customers to carry out terrorism vulnerability assessments. The initial assessments have been largely completed. The EPA has also drawn up plans for a monitoring system, called WaterSentinel, to detect chemicals and biological agents in our water supplies. Such a system is every bit as important as monitors for airborne pathogens, but the technology is still lagging, and funding has been slow. The plan has yet to be implemented.[4]

The classic fear most of us have of terrorists emptying vials of deadly poison into city reservoirs turns out not to be terribly likely, mainly because of the enormous dilution factor involved. The amounts of chemical that would have to be added to even a modest-size reservoir to reach a chemical concentration toxic to humans in a glass of drinking water approaches boxcar-size loads, making this method of attack highly impractical. While the same is generally true of biological agents, many microbes can not only survive in water, but actually multiply using nutrients commonly found in most water sources.

We know what can happen when the water supply to metropolitan areas is contaminated with human pathogens, through numerous incidents of accidental intrusion of pathogens into the water supply. The city of Milwaukee, Wisconsin, uses water from Lake Michigan as one source of its drinking water supply. Water from the lake is passed through a series of purification steps at two city water purification plants before entering the drinking water distribution system. In late March 1993, citizens and some city officials began noticing an increasing turbidity of tap water, particularly on the city's south side.

In early April, hospitals and clinics began reporting increasing numbers of people coming in with gastrointestinal problems, including severe cramps, persistent diarrhea, and vomiting. It always takes a day or two to realize there is a problem, because some level of these symptoms are normally seen on a daily basis in large cities. It takes another day or two to figure out what the problem is. In this case, analysis of patient stool samples indicated the presence of massive amounts of the intestinal parasite *Cryptosporidium*, and its source was traced to the city's drinking water.

The initial water filtration system at one of the two city plants had failed. The plant was possibly also affected by an unusual amount of runoff from rivers and streams into the lake, dragging with it larger than normal amounts of waste products, especially from livestock fields. Ultimately, over 400,000 people were made ill, roughly half of those who had imbibed local water. (Many healthy people remain unaffected by *Cryptosporidium;* older people, sick people, and those with impaired immune systems are at highest risk.) Over 4,000 people required some hospital treatment, and 54 people died. More than 700,000 work days were lost, and the cost to the community has been estimated at up to $100,000,000.

In another natural disease outbreak, traceable entirely to contaminated runoff water, the culprit was the bacterium *E. coli* 0157:H7, an agent on the CDC "B list" of potential biowarfare pathogens (Table 3.1), and *Campylobacter.* Both cause symptoms similar to the parasite *Cryptosporidium*, but the diarrhea caused by the *E. coli* is usually bloody. In the five-day period May 8–12, 2000, the small rural town of Walkerton in Ontario, Canada, experienced five and a half inches of rain, which came after a long period of drought. Three and a half inches fell on May 12 alone. This combination of events was described as a "once-in-a-hundred-years event."

Walkerton obtains its water from a series of local wells. These wells rely on natural filtration of groundwater in underlying aquifers, and chlorination of surface water entering the wells. There are no formal purification plants operating. During the excessive rainfall, manure from nearby fields was carried by surface water into the wells, overwhelming the ability of the chlorine to deal with the accompanying bacteria. *E. coli* 01577:H7 is common in animal manure, and a major concern about runoff from grazing fields into water systems. The result was that many hundreds of people were made ill, and seven died. Estimates of the overall costs of this incident also approach $100,000,000.

These cases make clear that we would indeed be vulnerable to bioterrorism through our water systems. There are several points

BOX 6.1

INTRODUCTION OF BIOLOGICAL AGENTS INTO THE WATER SUPPLY

Primary sources	Lakes, rivers, streams, dams
Intake points	Points of entrance of primary source water into purification plants; shared water wells
Water distribution systems	Water pipes running throughout cities
Stored purified water	Reservoirs, cisterns, water tanks
Commercial water	Bottling plants, carboys, bottles
Recreational waters	Lakes, swimming pools

at which pathogenic agents could be introduced by terrorists into our water (Box 6.1). The most likely targets for deliberate water contamination would be water intake points and water distribution systems. Recreational waters are also of some concern for pathogens that can cling to skin and hair.

HOW CAN WE PROTECT OURSELVES?

The openness of our croplands and the tremendous concentration of our livestock (80 percent of our cattle at any one time are concentrated in a half dozen or so feedlots) make agriculture a prime target for terrorists. Many crops are similarly concentrated in a limited number of huge processing plants. Weaponization and delivery of crop and livestock pathogens requires considerably less sophistication than biological attacks on human targets living in metropolitan areas, and pose no risk for those who use them. We also know that terrorists linked to Al-Qaeda have looked very closely at our water supply and delivery systems. Documents and computer drives recovered from Al-Qaeda sites in Afghanistan contained a surprising number of detailed maps and operating procedures for water systems in several American cities.

And we need not worry only about foreign terrorists. Some of the more militant domestic groups opposed to the slaughter of animals for food, or to the increasing cultivation of genetically modified crops, could turn to terrorism to force their agendas onto the table of national discussion. In a sort of "covert agrowarfare," both the United States and the United Kingdom have invested in programs to develop bioweapons capable of taking out specific crops in other countries—specifically coca plants in South America and poppy plants in the Middle East.

As with bioterrorism against humans, federal authorities have taken a number of steps to improve our ability to defend ourselves against agricultural terrorism, whatever the source.

BETTER CONTROL OF POTENTIAL AGROTERRORISM PATHOGENS

The National Research Council concluded, in a 2002 report titled *Countering Agricultural Bioterrorism*, that the threat of terrorism against U.S. animals, crops, and the water supply is very real, because of both the social disruption it could cause and the economic damage it could do. But this report also recognized that it is not feasible to make and have on hand the scientific tools to counter every threat. On the other hand, as with human pathogens, we can identify those plant and animal pathogens that have the potential to cause the greatest harm to American agriculture and do everything possible to make sure these agents do not fall into the wrong hands.

The Department of Health and Human Services, together with the U. S. Department of Agriculture (USDA), has drawn up a "select list" of agricultural pathogens that are to be kept under extremely strict control. All of the restrictions placed on human pathogens in terms of licensing of distributors and users and tracking and reporting of all select agents also apply to the designated agricultural agents. These are agents which, if they escaped uncontrolled into the agricultural environment, could cause serious damage to the domestic food supply and to America's ability to sell food abroad. Some of these agents have been designated by the CDC and the USDA as potential agroterrorism agents (chapter 7).

THE STRATEGIC PARTNERSHIP PROGRAM

The Department of Homeland Security, the Department of Agriculture, the FDA, and the FBI joined together in 2005 to create the Strategic Partnership Program Agroterrorism Initiative. Teams representing each of these agencies travel around the country meeting with state and county representatives, as well as individual private agricultural enterprises, to identify where the vulnerabilities in our agricultural infrastructures lie and how we can best ameliorate those vulnerabilities. Questions for discussion include how best to detect actual or potential agroterrorist attacks and how to deal with them once they occur. For example, from economic, political, and social points of view, is mass destruction of livestock in outwardly growing "rings" the most effective way to deal with something like a foot-and-mouth disease outbreak, or might there be a more imaginative way to use vaccines, perhaps in conjunction with mass destruction? Clarifying and strengthening the respective roles of state and federal governments, and providing closer communications between government at all levels and the private sector during emergencies, are also top priorities.

Individual acts of bioterrorism, whether directed at human or agricultural targets, are low-probability events. Nevertheless, any agroterrorist attack, if carefully planned—and if successful—could cause substantial damage that could take years to recover from. How much do we invest in protecting ourselves? How much to protect our livestock? Our crops? That is a very difficult policy question, but to the extent we decide to invest in bioterrorism defense, agroterrorism certainly cannot be ignored.

CHAPTER 7

AMERICA RESPONDS

IN 1989, PARTLY AS A RESULT OF AMERICA'S PARTICIPATION in the 1972 International Biological Weapons Convention, but also in response to growing concerns about terrorist use of biological weapons, Congress passed the first of a series of bioterrorism acts, this one entitled the Biological Weapons Antiterrorism Act of 1989. This Act spelled out the first restrictions on access to biological weapon agents by individuals:

> Sec 175a. Whoever knowingly develops, produces, stockpiles, transfers, acquires, retains, or possesses any biological agent, toxin, or delivery system for use as a weapon, or knowingly assists a foreign state or any organization to do so, shall be fined under this title, or imprisoned for life or any term of years, or both.

This is the law under which members of the Minnesota Patriots Council were tried and sentenced for manufacturing and possessing ricin, even though they never used it. Subsequent sections of this Act provide a legal basis for searches and seizures of biological agents and spell out conditions for use of such agents by scientists engaged in studies within recognized research institutions with appropriate institutional and granting agency oversight. Provisions of the Act were clearly targeted toward persons buying, making, or possessing dangerous human pathogens that have "no apparent justification for prophylactic, protective, or other peaceful purposes."

The next significant piece of legislation to appear had the grim title Antiterrorism and Effective Death Penalty Act of 1996. It focused primarily on strengthening and streamlining federal law governing imposition of the death penalty in certain terrorism cases. It was spurred largely by the Oklahoma City bombing and the first World Trade Centers attack in 1993, and was given a further boost by the Aum Shinrikyo incidents in 1995. The Act also made it easier for victims of terrorist acts to be compensated for their losses, and authorized them to sue foreign governments in federal courts if those governments could be shown to have supported terrorists who harmed them.

Title V of this Act focused additional attention on transfer of and trafficking in biological agents that could be used by terrorists, and was to a significant degree impelled by the activities of the Rajneeshees, the Minnesota Patriots Council, and Larry Wayne Harris (a credit he will doubtless add to his resumé!). In the final stages of drawing up the Act, unfolding knowledge of Aum Shinrikyo's attempts to create bioweapons added to a sense of urgency in Congress to create stronger controls. The Department of Health and Human Services (HHS) was directed to draw up a list of exactly which biological agents were to be prohibited by federal law from possession by unauthorized individuals. This resulted in the HHS "Select Agents" list (see below).

The ink had hardly dried on the 1996 Act, and its full implementation was still underway, when the United States was struck by the attack on the World Trade Center and the subsequent Amerithrax incidents of 2001. The General Accounting Office, in a report prepared for Senator Bill Frist, concluded that this episode in our history with bioterrorism revealed glaring weaknesses in the state of preparedness of our emergency and public health systems to deal with even moderate-scale bioterrorist attacks. In response, a large-scale, bipartisan effort initiated in 1999 was brought rapidly to completion, in the form of the Public Health Security and Bioterrorism Preparedness and Response Act of 2002 (which from here on out we will refer to as the 2002 Bioterrorism Act). Intended to be amended on a regular, ongoing basis, it provides a blueprint for upgrading American preparedness to deal with catastrophic biomedical emergencies. It incorporated, modified, and extended many of the features of the 1989 and 1996 Acts.

BOX 7.1

KEY PROVISIONS OF THE PUBLIC HEALTH SECURITY AND

BIOTERRORISM PREPAREDNESS AND RESPONSE ACT OF 2002

Title I   Calls for the upgrading of public health services
               at every level to deal with catastrophic health
               emergencies. Decrees creation of detailed state
               and local response plans, training of emergency
               response personnel, improvement of hospitals to
               deal with catastrophic events, creation of national
               stockpiles of medicines and vaccines.

Title II   Defines biological agents and toxins deemed to
               pose a particular threat to humans. Sets strict
               regulations for controlling access to these agents
               and toxins, requires licensing of both suppliers and
               users, and establishes criminal penalties for their
               unauthorized possession or use.

Title III   Defines steps to be taken to protect domestic food
               and drug supplies.

Title IV   Describes steps to be taken to secure the domestic
               water supply.

The 2002 Bioterrorism Act is over 100 pages long[1]; its most important features are shown in Box 7.1. Although aimed primarily at preparing public health services to deal with a bioterrorist attack, part of selling it to Congress and the public was that it would also help prepare for any large-scale infectious disease crisis. The vast majority of cities and counties in America are equally underprepared for any major flu pandemic, and H5N1 was already on the radar screen.

Title II of this Act incorporates and extends the Select List of pathogenic agents deemed likely to be involved in catastrophic health incidents (Table 7.1), and refines the provisions covering their use. Since the 2002 Act came into effect, only certain licensed laboratories are able to distribute materials from the Select List, and only licensed end users are authorized to purchase them. Companies like ATCC (American Type Culture

TABLE 7.1   The Department of Health and Human Services/USDA Select
List of Agents and Toxins

Abrin	Lassa fever virus
*Bacillus anthracis**	Marburg virus
*Botulinum neurotoxins**	Monkeypox virus
*Brucella abortus**	Nipah virus*
*Brucella melitensis**	Ricin
*Brucella suis**	*Rickettsia prowazekii*
*Burkholderia mallei**	*Rickettsia rickettsii*
*Burkholderia pseudomallei**	Rift Valley fever virus*
*Clostridium botulinum**	Saxitoxin
Clostridium perfringens toxin*	Shigatoxin*
*Coccidioides immitis**	South American hemorrhagic fever
Coccidioides posadasii	viruses (5 types)
Conotoxin	Staphylococcal enterotoxins*
*Coxiella burnetii**	T-2 toxin*
Crimean-Congo hemorrhagic	Tetrodotoxin
fever virus	Tick-borne encephalitis viruses
Diacetoxyscirpenol	(5 types)
Eastern equine encephalitis virus*	Variola major and Variola minor
Ebola virus	Venezuelan equine encephalitis
*Francisella tularensis**	virus
Hendra virus*	*Yersinia pestis*
Herpes B virus	
Influenza virus (1918 variant)	

*Bold type: agent also appears on the CDC list of most dangerous potential bioterrorism
agents (chapter 3).*
Agent also of concern for agricultural terrorism (chapter 6).

Collection), one of the largest U.S. suppliers of both microbial
and animal cell lines, is an example of a company certified by the
CDC for transferring Select List agents to authorized researchers. In order to receive CDC certification to transfer these materials, ATCC had to submit extensive documentation about how
it records and stores these materials internally, who has access
to them, and security measures in place to prevent unauthorized movements both within and outside of ATCC. Each time

someone tries to purchase a Select List agent, ATCC must file a special form with the CDC, and the individual requesting the agent must do the same. CDC then verifies that this individual has previously filed all of the necessary paperwork to receive the requested agent and notifies ATCC of its decision regarding the requested shipment. The possibility that someone other than a vetted researcher in an authorized institution could obtain a Select List agent from ATCC is just about zero.

The whereabouts of every trace of these agents must now be recorded and a paper trail created following their movement within both the transferring company and the researcher's home institution. Possession of, or trafficking for any purpose in, these agents and toxins outside the proscriptions of the Act is subject to stiff criminal penalties. Some of these agents and toxins are on the Category A list of the Centers for Disease Control and Prevention, as materials of particular concern in bioterrorism (chapter 3).

The cases of bioterrorism and biocrime we examined in chapter 2 are just a small part of the many cases of terror, using bombs and gas as well as biological agents, that have become part of the background noise of modern political and social life. Bioterrorism is admittedly part of that noise, and has the potential to do substantial damage. But what will be the likely source of that bioterror? Will it come from without, like the 1993 and 2001 World Trade Center attacks? Or will it come from within, like most of the cases we explored earlier? That is what we will spend much of the rest of this book examining. But to paraphrase a famous cartoon character of years past, 'We have met the enemy— and so far, he is mostly us!'

CREATING AN INFRASTRUCTURE: THE STRATEGIC NATIONAL STOCKPILE AND PROJECT BIOSHIELD

The Strategic National Stockpile

As part of strengthening our public health emergency response systems, Congress asked HHS in 1999 to create the National Pharmaceutical Stockpile to speed medical supplies and materials to the sites of major health emergencies anywhere in the United States within twelve hours. The response to this call was rapid, and indeed CDC, a branch of HHS, was able to do just that immediately after the World Trade Center attacks of 2001. A fifty-ton, pre-packed shipment of medical supplies reached New York City within seven hours of being called upon, in one of the few non-military aircraft flying in American airspace that night. This same Stockpile was called upon again in the Amerithrax attacks just weeks later, when a hundred cases of medical supplies for dealing with anthrax were shipped to Florida.

In March of 2003, the National Pharmaceutical Stockpile was expanded and upgraded to become the Strategic National Stockpile (SNS),[2] and the Department of Homeland Security became a partner in its management. The official mission of SNS is to "ensure the availability and rapid deployment of life-saving pharmaceuticals, antidotes, other medical supplies, and equipment necessary to counter the effects of nerve agents, biological pathogens, and chemical agents." The budget for SNS reflected its recognized value and expanded role. The National Pharmaceutical Stockpile was initially funded at $50,000,000 per year. The current annual budget for SNS is around ten times that.

CDC has created ready-to-go "Push Packages," pre-packed cargo containers that can be transported by surface or air to any site in the United States within the twelve-hour delivery target. These contain a wide array of medicines and supplies that could be used in a bioterrorist attack, but to some extent in a number of other health emergencies as well. Twelve such Push Packages, each consisting of 124 separate cargo containers weighing a total of 94,000 pounds, are stashed in climate-controlled warehouses at strategic (and secret) locations around the country. Their medicine contents are rotated on a regular basis to assure freshness. Arrangements for their shipment have been made by special contract with both United Parcel Service and Federal Express. A Push Package will fit into one wide-bodied cargo plane or seven large 18-wheeler cargo vans. The CDC asks that all states have a dedicated, climate-controlled facility of 12,000 square feet to receive and temporarily store Push Package materials, and specific plans for their utilization.

Push Packages also come with a team of technicians who are thoroughly familiar with the contents and organization of each of the cargo containers, to assure rapid retrieval and efficient deployment of emergency supplies. Once a Push Package arrives, however, its contents become state property, and state and local health authorities direct their use and distribution. It is expected that at the end of an emergency using Push Packages, any left-over medicines and supplies will be melded into the state's health care system.

Push Packages are built primarily for speed of response. Early in an emergency, the exact scope and nature of the situation may not be entirely apparent. Subsequent aid, once authorities have a clear picture of the emergency and their needs for dealing with it, can be more tailored. In most cases, this aid may come directly from drug companies and manufacturers that are partnered into the SNS program and maintain their own "vendor-managed inventories" (VMI) for emergency distribution. Company X, for example, will guarantee always to have on hand for immediate shipment 250,000 doses of vaccine A or antibiotic B. The advantage of VMI is that the government does not have to maintain huge stockpiles of drugs with short shelf lives. Drugs rotated out of storage would have to be destroyed and replaced. This problem is bypassed through VMI by drawing on the manufacturer's own constantly replenished inventory.

In those cases where the exact nature of a health emergency is known early on, the SNS response may bypass Push Packages altogether and go directly to VMI. For example, supplies of cipro-floxacin (Cipro) were supplied directly and quickly through the VMI component of SNS in response to the Amerithrax attacks. More than 30,000 people received Cipro or equivalent antibiotics. The relatively low numbers of casualties in that situation are thought by many to be due to the immediate availability and essentially unlimited amounts of these drugs.

Like many of America's countermeasures against bioterrorism, SNS is a dual-purpose enterprise. It strengthens our preparedness for natural outbreaks of biological catastrophe as well as man-made incidents. In early 2006, for example, HHS authorized the purchase of 2.2 million and 3.8 million doses, respectively, of the antiviral drugs zanamivir and Tamiflu, bringing to

26 million doses the amount of antiviral drugs in SNS. Some of the antiviral drugs being stocked are also useful against viral pathogens on the CDC's A and B lists, but in any case we will have an increasingly broad spectrum of such drugs to draw upon in any major health emergency.

FEEDING THE BEAST: PROJECT BIOSHIELD

Project BioShield, signed into law in July 2004, was created to ensure an adequate flow of new and existing drugs, antiserums, and vaccines into the SNS armory, particularly those used for agents on the CDC A list, as well as medicines to protect against chemical and radiological attacks. SNS would also be the natural repository for drugs or vaccines intended to prepare for an avian influenza pandemic.

BioShield establishes funding for two main functions: research into the development of new vaccines and drugs to meet the threat of anticipated catastrophic health emergencies, and a standing cash source to purchase these new drugs and to replace existing SNS Push Package medicines and materials as they expire. Like SNS itself, this fund is overseen jointly by the secretaries of HHS and Homeland Security. Research grants processed through BioShield receive expedited review and decisions about funding.

The Food and Drug Administration (FDA) is also a partner in BioShield, helping to identify areas in which more research is needed. The FDA interacts extensively with the pharmaceutical industry as the chief overseer of clinical trials and is well positioned to help expedite research and testing of new drugs and materials destined for the SNS; it has the authority to put promising drugs and vaccines on a developmental fast track. The FDA would also play a key role in determining whether the possible benefits of using a particular drug or vaccine, which might still be in the developmental process during a national health emergency, could outweigh the risks posed by the emergency itself.

BioShield has been criticized by industry partners, and by some in government, for a provision that requires companies to actually produce an effective vaccine or drug before receiving any compensation from BioShield. Most of the research demonstrating efficacy of a drug or vaccine on a laboratory scale has usually been done in university research laboratories, paid for by federal research grants. But companies must take the laboratory findings and scale them up to industrial standards, as well as meeting tough FDA safety and efficacy guidelines. This can be an expensive process, and technical problems can be encountered in the scale-up which cannot be predicted from the preparatory research done in universities. This has caused some major drug companies to decline to engage in BioShield projects, leaving smaller and less experienced companies to fill the need.

This aspect of BioShield is of continuing concern to HHS, the FDA, and company leaders. As Congress adjourned at the end of 2006, a bill was passed (the Pandemic and All-Hazards Preparedness Act) providing $1,000,000,000 over two years to aid companies in early-stage research on needed drugs and vaccines. Use of these funds will be overseen by a new branch of NIH, the Biomedical Advanced Research and Development Authority (BARDA). This Act also clarifies the lead role of HHS, to the exclusion of Homeland Security, in the health and medical aspects of all public health emergencies.

STANDING GUARD: THE BIOWATCH AND BIOSENSE PROGRAMS

Among the lessons gleaned from the Amerithrax incidents was the need to identify when a bioterrorist attack has occurred as early as possible. Most pathogens have a significant incubation period inside a host before disease symptoms appear. By that time, the pathogen has expanded billions of times within the body, and the resultant disease may be difficult to control.

While we may never know the exact sequence of events during Amerithrax, it was probably at least a week after the anthrax spores were disseminated that the first infected individuals began showing up at local hospitals. Even then, unfamiliarity of medical personnel with anthrax resulted in additional delays before correct diagnoses were made and appropriate treatments initiated. These delays in combination likely resulted in some deaths that could have been prevented if action had been initiated sooner. If doctors and hospitals had been warned to be on the lookout for anthrax cases, with symptoms clearly described, earlier detection of active infections could have been made. Prompt administration of antibiotics such as Cipro or other medications—even on a prophylactic basis—could have slowed the spread of disease within individuals and reduced fatalities. SNS resources could have been mobilized more rapidly. That is the rationale behind the BioWatch Program.

BioWatch, established in 2003, is essentially a network of early warning biosensors under the overall supervision of the Department of Homeland Security. The network consists of three main components: sampling, analysis, and response.[3] Each of these components is overseen by a separate agency. Sampling is overseen by the Environmental Protection Agency (EPA), which has placed filters capable of trapping airborne pathogens into its already established system for monitoring air quality. Analysis of these filters is carried out every twenty-four hours in regional testing laboratories[4] under supervision of the CDC. During a given twenty-four hour period, filters within the sensors are imprinted with the time and automatically replaced every hour, to pinpoint the exact time of an attack.

Details of the BioWatch program are understandably vague, for security reasons. The initial round of installation of sensors included approximately thirty cities, with an assumed eventual goal of covering perhaps a hundred metropolitan areas. While some sensors are associated with existing EPA air quality monitoring stations, it is thought that many others are not. Siting of permanent sensors is a highly sophisticated science, utilizing satellite imagery and detailed climatic data from the U.S. Weather Service. So-called "special event" sensors are more mobile, and can be tailored to specific sites. Mobile sensors were strategically placed in venues for the 2002 Winter Olympics in Salt Lake City, for example. Special sensors have also been set up in major postal handling facilities.

BioWatch sensors have triggered several alarms to date, in San Diego, Washington, D.C. and Houston. The Houston and Washington incidents involved *F. tularensis,* the bacterium that causes tularemia; the alarm in San Diego was caused by *Brucella* sp. Both pathogens are present to some extent in the natural environment. In all three cases, the levels detected were below those officially set to trigger a full response, but local health authorities were alerted to be on the lookout for specific symptoms. Exhaustive analysis of all data suggested that these three occurrences were natural rather than the result of any overt action.

The BioWatch Program is not without its critics. It has been pointed out that it would have had no impact on the Amerithrax attacks, because the spores were all released within buildings, and would not have reached EPA sensors mounted in the affected metropolitan areas. Defenders counter that at the very least, knowledge that such detection systems exist could lessen the likelihood that terrorists would mount a major outdoor attack in large cities.

Concern has also been expressed about the cost of equipping and maintaining a large network of sensing stations, especially in the labor-intensive data analysis phase. The Department of Energy is currently working on a completely autonomous pathogen detection system that could drive down the costs of analysis, and reduce even further the incidence of false-positive and false-negative incidents. But BioWatch is not one of the "dual-purpose" components of recent improvements in American public health preparedness. It is aimed solely at managing a bioterrorist attack.

If bioterrorism is a low-probability/high-consequence event, how much is it worth to buy the extra security that BioWatch costs U.S. taxpayers? Current estimates are that it costs about one to two million dollars to install a BioWatch system in each metropolitan area, and about two million dollars a year in each area to maintain the system and process samples. The benefits of early detection depend on the pathogen used in an attack. Most cost-benefit estimates have been based on analyses of an anthrax attack, because that is what we know most about. But anthrax, while deadly, is not infectious. The benefit of early detection in the case of an infectious agent like smallpox could be significant.

The BioSense Program is a still-evolving program with the objective of using computer-based systems for analyzing reports of health-threat incidents from major metropolitan centers across the United States. The purpose is to provide the earliest possible alarm that unusual disease patterns may be emerging, whether they result from terrorism or from natural outbreaks. Participating medical centers will share their own computerized records of illness reports with the BioSense Program, which will develop new electronic tools for spotting potentially important clusters of disease incidents that require further and immediate analysis. Many of the software programs to do this are still in a developmental stage, but when fully operational are expected to add critical hours to days in the public health response to emerging outbreaks.

The federal government has established a set of goals that states should meet in order to consider themselves prepared to deal with catastrophic health emergencies. It must be admitted that many states are still struggling to meet these goals; as of 2006, only one (Oklahoma) had met all ten preparedness benchmarks (Box 7.2).[5] Still, the United States has made impressive progress in the past half-dozen years in building a stockpile of drugs, vaccines, and equipment that would greatly limit the damage terrorists might inflict on us with bioweapons. This doesn't mean we are completely immune to bioterrorist attacks, but the knowledge that the damage done by such attacks would be considerably less than might have been hoped for just a few years ago, together with the technical difficulties in producing an effective bioweapon, could be a major deterrent for many terrorist groups.

Total federal spending on programs to defend us against bioterrorism are shown in Figure 7.1.[6] The amounts indicated do not include spending for Project BioShield, which totals about $3,300,000,000 across the period shown. If the 2007 and 2008 amounts hold, the United States will have spent about $45,000,000,000 on biodefense since 2001.

THE NATIONAL STRATEGY FOR PANDEMIC INFLUENZA, 2006

It's been a favorite mantra over the past two decades among those urging America to move forward more forcefully in its preparation for a bioterrorist attack: "It's not a question of *if*," they say, "but

BOX 7.2

PREPAREDNESS SCORES (BY STATE)

4	5	6	7	8	9	10
California	Alaska	Colorado	Delaware	Alabama	Kansas	Oklahoma
Iowa	Arizona	Indiana	Florida	Kentucky		
Maryland	Arkansas	Louisiana	Georgia	Michigan		
New Jersey	Connecticut	Massachusetts	Hawaii	Missouri		
	D.C.	Mississippi	Idaho	Montana		
	Maine	Nevada	Illinois	Nebraska		
	Ohio	New Mexico	Minnesota	South Dakota		
	Pennsylvania	No. Carolina	New Hamp.	Texas		
	So. Carolina	Oregon	New York	Virginia		
		Rhode Island	No. Dakota	Washington		
		Utah	Tennessee	West Virginia		
		Vermont		Wyoming		
		Wisconsin				

States are listed under the total number of ten federal benchmarks for bioterrorism preparedness they have met. For a complete discussion of the ten federal benchmarks, see note 5.

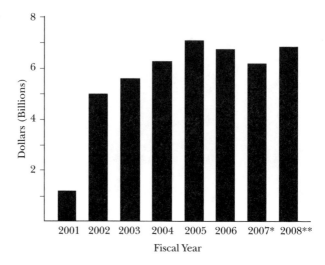

FIGURE 7.1 *Biodefense spending by fiscal year. *Estimated. **Requested.*
Excludes funding for BioShield.

when." That could conceivably be true with regard to bioterrorism; it is an absolute certainty with natural pandemics. As of April 2007, H5N1 had appeared in wild or domestic birds in over fifty countries, including the United States (in wild swans in Michigan in August 2006, and in wild ducks in Pennsylvania and Maryland in September of that year). Hundreds of millions of birds worldwide have died from the virus, or been culled. The virus is so widespread in Southeast Asia that it is now considered endemic there. As the number of birds infected with H5N1 increases, the odds of a mutation occurring that makes either bird-human or human-human transmission easier increases accordingly. And that spells pandemic.

The past record is clear: pandemics are simply a fact of life that we must adjust to, like death and taxes—and, yes, global warming. We cannot say the next pandemic will be caused by H5N1, although for now that seems the most likely candidate. Nor can we say when the next pandemic will be, or how severe, any more than we can say when the next bioterrorist event might be or how much damage it could cause. But if it's true that we need a plan to deal with bioterrorism, it is even more true that we'd better have a plan for the next pandemic.

True, the several Acts we have talked about so far in this chapter also help us prepare to defend ourselves against a pandemic. But the bulk of the effort and money generated by these Acts have been directed specifically against the threat of bioterrorism. Many in this country have long argued that this is a bit like the tail wagging the dog. Bioterrorism may be a threat, they say, but one of low probability and indeterminable cost. Natural pandemics are a slam dunk. And now those who think pandemics should have a plan of their own to line up behind have one. Its supporters are confident that actions taken under its prescriptions will more than adequately prepare us for a bioterrorism attack. Finally, they say, the dog is wagging the tail.

The National Strategy for Pandemic Influenza (NSPI), drawn up by the Homeland Security Council, was presented to the public in several stages beginning in November 2005.[7] NSPI takes a wider aim than the previous bioterrorism acts, as articulated in the opening document: a pandemic health crisis will necessitate a strategy that

> extends well beyond health and medical boundaries, to include the maintenance of critical infrastructure, private sector activities, the movement of goods and services across the globe, and economic and security considerations.... [NSPI] guides our preparedness and response to an influenza pandemic with the intent of (1) stopping, slowing or otherwise limiting the spread of a pandemic to the United States; (2) limiting the domestic spread of a pandemic, and mitigating disease, suffering and death; and (3) sustaining infrastructure and mitigating impact to the economy and the functioning of society.

According to the official Implementation Plan for NSPI, presented in May 2006, NSPI

> provides a high-level overview of the approach the Federal Government will take to prepare for and respond to a pandemic, and articulates expectations of non-Federal entities to prepare themselves and their communities.

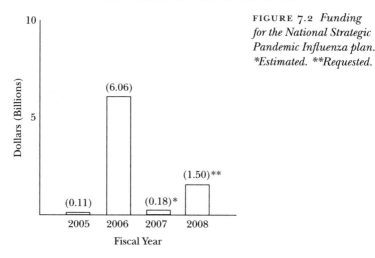

FIGURE 7.2 *Funding for the National Strategic Pandemic Influenza plan. *Estimated. **Requested.*

President Bush asked Congress for $7,100,000,000 over several years to fund the aims of NSPI; it appears that Congress may supply somewhat more, and about 90 percent of that will be allocated to HHS (Figure 7.2). The Pandemic and All-Hazards Preparedness Act of 2006, referred to earlier, requires HHS to set influenza pandemic public health preparedness standards for states, and defines penalties for failure of states to do so. Oversight of the federal effort to implement NSPI will rest jointly with the Department of Homeland Security and HHS. DHS will coordinate federal operations, resources, and communications with state, local, and private entities. The Secretary of HHS will be the federal spokesperson on all public health and medical issues.

NSPI and its Implementation Plan define three major pillars of America's response to the threat of a catastrophic influenza pandemic. These do not define three sequential actions we must pursue to manage a pandemic crisis, but rather three interacting concerns that affect each action we take:

• Preparedness and communication—Activities that should be undertaken in advance of a pandemic to ensure we are as prepared as we can be, and clear communication of the roles and responsibilities expected of all levels of government and individuals in managing the crisis;

- Surveillance and detection—Domestic and international monitoring systems to provide continual tracking of potential pandemic viruses, so that we have the earliest possible warning of a pandemic that may reach the United States;
- Response and containment—Actions to limit the spread of a pandemic once it is officially declared, in order to mitigate the health, social, and economic consequences to the American people.

The Implementation Plan sets forth a number of steps the United States must take to meet these goals.

PREPAREDNESS AND COMMUNICATION

Develop and Stockpile Vaccines

Currently there are two substrains of H5N1 wreaking havoc among birds. A limited amount of vaccine has been prepared against one of them, as discussed in the last chapter, and is currently being stockpiled. Work on a vaccine for the second substrain is proceeding apace. Both of these will be "best-guess" pre-pandemic vaccines. We won't know if they are effective against the pandemic H5N1 variant until that variant actually arrives. The goal is to have enough pre-pandemic vaccine to immunize 20,000,000 people. The hope is that pre-pandemic vaccines will be useful in slowing spread of a pandemic until a vaccine specifically tailored to the pandemic agent can be prepared.

In addition, the federal government will provide financial assistance to upgrade existing vaccine production facilities in the United States, and to build new ones if necessary. The government, through NSPI, will also fund research into more rapid and effective procedures for producing vaccines. In the 1970s, there were two dozen vaccine manufacturers from which America could buy flu vaccines; today there are five. A large proportion of our seasonal flu vaccines come from other countries. In the event of a worldwide flu pandemic, we may not be able to rely on those resources. Those of us over sixty (the primary demographic group for flu vaccination) remember what happened in 2004, when one of the overseas facilities had to shut down; the United States had less than 40 percent of the needed doses of flu vaccine that year, and millions went without a seasonal flu shot.

STOCKPILE ANTIVIRAL DRUGS AND OTHER SUPPLIES AND EQUIPMENT

The domestic goal is to have enough drugs such as Tamiflu or Relenza in federal and state repositories to treat 25 percent of the population, or about 75,000,000 people. As of spring 2006, enough courses of Tamiflu to treat about 20,000,000 people had been stockpiled.[8] Antivirals will also be positioned, under international control, near pandemic high-risk countries. Emphasis has been on Tamiflu, since this drug can be used to prevent infections in persons at high risk at the site of an outbreak as well as treat those already infected. Influenza A viruses are rapid mutators, and substrains resistant to any particular drug could arise under pressure from that drug. NSPI thus also supports research into developing new antiviral drugs, which are usually less easy to escape from by mutation on the part of the virus.

Finally, NSPI will provide funds to purchase up to 6,000 new ventilators, 50,000,000 surgical face masks, and 50,000,000 N95 respirators[9] for domestic use. Many more ventilators would be needed to handle even a moderate pandemic; suffocation during pneumonia was the most common cause of death during the 1918 pandemic. But it's a start.

COORDINATE FEDERAL, LOCAL, AND STATE RESPONSES

The most important way that a pandemic will differ from a bioterrorist attack is that the latter, like a major earthquake or hurricane, will be a localized disaster, with rapidly mobilized help rushing to the site from all around the country. In the United States, the full force of protections such as the Strategic National

Stockpile, VMI, FEMA, and other federal and state resources can be focused rapidly on the site of the attack. A pandemic, by definition, will spread rapidly to many sites throughout the world and within the United States itself, with both domestic and foreign "hot spots" shifting constantly. The NSPI emphasizes that a pandemic will be more like a full-scale war than a terrorist attack, and goes on to state:

> in a pandemic, conditions would make the sharing of resources and burdens even more difficult.... in the event of multiple simultaneous outbreaks, there may be insufficient medical resources or personnel to augment local capabilities.

Thus authorities are already warning that every level of our polity—individuals, families, cities, counties, and states—must develop self-sufficiency plans to protect and sustain themselves as long as possible. In a serious pandemic, there may be little or no help flooding in from anywhere for quite some time. To this end, NSPI is organizing a series of Dark Winter–like exercises in every state to test preparedness for a pandemic, and states are expected to conduct similar exercises at county and city levels.

Enhance Communications and Public Awareness

As in a bioterrorist attack, but on a much broader scale, it will be extremely important that accurate information and advice that the public trusts and follows be made available. But unlike a bioterrorist attack, a pandemic will stretch over a much longer period of time, on the order of months at the least, and up to a year or more. People who give out information or advice should themselves be experts in the areas they are discussing, and should be retained throughout the crisis for purposes of public continuity. Inadvertent misrepresentation of facts in such situations can be disastrous. Information to be distributed must take into account the changing ways in which people access information in an information technology society. It will also be important to provide U.S. citizens with accurate, up-to-date travel advisories.

SURVEILLANCE AND DETECTION

Detecting an Emerging Pandemic

A key element underlying the NSPI is to identify as early as possible, at the initial site of a potential pandemic anywhere in the world, the precise identity of the underlying virus. Speed is of the essence: all three twentieth-century flu pandemics, although of widely varying severity, engulfed the entire globe in just a few months. Identification of a virus can be done rather quickly, but it may take a week or so for experts to be convinced that the viral variant has all of the elements necessary to cause a pandemic: the ability to pass readily from animals to humans, the lack of any immunity to the variant in the human population, and ready transmission of the variant between human beings. Absolute certainty before sounding an alarm is critical: the last thing anyone wants, for as long as such a crisis looms over us, is a false alarm, leading to a massive worldwide mobilization that has to be withdrawn. This would result in a loss of confidence within the public and a more tentative response when a real pandemic emerges.

The United States will work with other countries and with the WHO and other international agencies to track the wanderings of H5N1 or other flu A variants as they travel the world with migratory birds. To this end also, the United States will greatly increase monitoring of migratory birds as they pass through its own territory. As we have seen, birds harboring H5N1 have already been detected in the United States.

Identifying and Treating Infected Individuals

Once a pandemic is underway, speed of identification of infected individuals is of the utmost importance. In the case of influenza, infected persons can begin spreading the virus to others in as little as forty-eight hours after infection. Estimates are that the number of persons infected will likely double every three days in crowded locations. There are test kits available to public health officials that can analyze fluid samples for the presence of influenza virus in about an hour. Unfortunately they cannot as yet distinguish different variants. NSPI is supporting research into making these tests faster and more specific.

CONTROLLING THE PANDEMIC INTERNATIONALLY

It will be in the interest of the United States to work to prevent, and participate in controlling, a pandemic that may surface in another country, to prevent or at least slow its spread to this country. The United States, through NSPI, has already established an International Partnership on Avian Pandemic Influenza. Through the partnership, and in cooperation with international health agencies, the United States will assist countries deemed at high risk for a pandemic outbreak to develop state-of-the-art monitoring facilities and to improve veterinary facilities, among other things.

VACCINE PRODUCTION

One of the most important outcomes of early detection and definitive identification of a pandemic influenza variant will be the ability to begin immediately to design and produce a vaccine specifically tailored to control of this variant. All other responses at the beginning of a pandemic are aimed at slowing its spread by other means until such a vaccine is available. NSPI is thus also providing funds to enhance the ability of vaccine manufacturers anywhere in the world to respond with all possible speed once the variant is identified.

RESPONSE AND CONTAINMENT

Interestingly, the NSPI rests almost entirely on the assumption that a pandemic would arise outside the United States, and the response and containment plan is formulated accordingly. Much of the initial discussion in Chapter 5 of the Implementation Plan focuses on "securing the borders." But it is not at all inconceivable that the crucial mutation(s) required to create a worldwide pandemic could arise in wild or domestic birds right here in America. And while in many ways response and containment

actions would be the same, many of the social and economic dis-
locations resulting from a pandemic arising here would be different from one arising outside the United States. (For example, see
the Dark Winter scenario in chapter 1).

In terms of protecting human health, the Implementation
Plan rightly states that

> The cardinal determinants of the public health response to a
> pandemic will be its severity, the prompt implementation of local
> public health interventions, and the availability and efficacy of
> vaccine and antiviral medications. [Health measures will be]
> determined by the ability of the pandemic virus to cause severe
> morbidity and mortality, and the availability of and effectiveness
> of vaccines and antiviral drugs.

SURGE CAPACITY

One of the greatest challenges, for pandemics as well as for bioterrorist attacks involving contagious pathogens, is the problem
of medical surge capacity. In both cases, shifting either patients
or resources to other locations will be very difficult, often impossible. Again, local self-reliance will be the word of the day. Yet for
the past ten years we have been forced to slim our medical services to a bare minimum as we have attempted to control health
care costs. As a result we have almost no extra local capacity to
meet a nationwide health crisis.

NSPI will work with states and communities to develop plans to
deal with the surge in demand for health care personnel, supplies,
and equipment. States are urged to identify, certify, and register
in advance—now, not when a crisis arises—health care volunteers who can be called upon when the time comes. These could
include retired doctors and nurses, physician assistants, emergency medical technicians, and former military medics, among
others. The demand for hospital beds will certainly greatly out-
strip the few empty beds in most communities, and alternatives
must be planned for. Triage will likely be necessary to be sure persons receiving beds, medicines, and medical attention will benefit
from them. Demands for health care by those not affected by the
pandemic virus must somehow continue to be met.

We are continuing to fund bioterrorism preparedness at about $8,000,000,000 annually, the same rate at which it has been funded since 2004. NSPI has requested at least $7,000,000,000 to get its programs off the ground. The General Accounting Office can be expected to point out to our politicians that this may be an excellent opportunity to review the projects and funding in both programs and weed out needless duplication. The major focus of both remains the upgrading of public health and emergency medical services to deal with catastrophic health crises. We will soon have increased our protection against bioterror threats to a point where additional defenses may not justify the cost. We are far from that point with respect to dealing with a major pandemic, which has different requirements, but NSPI marks a real beginning.

CHAPTER 8

POLITICAL, LEGAL, AND SOCIAL ISSUES IN A NATIONAL HEALTH EMERGENCY

DURING THE DARK WINTER exercise of JUNE 2001, one of the questions the unscripted participants had to grapple with was: who is running the show in a nationwide health crisis? Is it the federal government, or individual states? If it's a joint effort, who's responsible (and has the authority) for what? Would it matter whether the crisis was precipitated by a natural outbreak of deadly disease or by a terrorist attack interpreted as a hostile action against the United States?

These are not questions that would arise in other modern industrialized nations, all of which have centralized, dominant governmental public health structures. It is a question in the United States of America in the twenty-first century because of our unique origin as a confederation of independent states that happened to occupy a common continental land mass. It could possibly become a question one day in the continually evolving European Union. They will have the advantage of having watched us struggle with questions of this type for over 200 years.

In the United States, it is generally agreed that preservation of the public health is a power granted to individual states, as an exercise of their constitutionally guaranteed policing powers.

But after the events of September 11, 2001, and especially in the wake of the subsequent postal anthrax attacks along the East Coast, both the executive and legislative branches of the federal government took steps to strengthen the U.S. public health response to catastrophic biological events, from whatever source. The Public Health Security and Bioterrorism Preparedness Act of 2002 (chapter 7) was one response, setting out what the federal government was prepared to do and what states were urged to do. This Act was given added impact by the creation of the Strategic National Stockpile and Project BioShield.

Reform was clearly needed. Existing public health laws in most states can charitably be described as a patchwork. Internally, many state laws evolved reactively, stitched together in an ad hoc fashion from responses to crises each state had experienced in its past. Each state's past experiences having differed, there was little conformity among states. Few state public health laws incorporate new health information and technology into their legal underpinnings until new situations requiring new approaches actually arise. Most also do not take into account changes in federal (and even their own state) laws dealing with issues such as privacy or patients' rights until a specific challenge is raised.

Even prior to September 11, there had been calls for a major overhaul of the public health system in the United States, which had been stretched to its limits dealing with the AIDS crisis beginning in the early 1980s. The prestigious Institute of Medicine in 1988 issued a major report calling for upgrades in the ability of public health systems to deal with large-scale health emergencies.[1] The federal government finally asked the Center for Law and the Public's Health, a consortium of faculty experts from Johns Hopkins University and Georgetown University, to draft a model for states' reactions to such crises, particularly with respect to laws that govern what state-level emergency responders and public health officials can and cannot do. Work on this document was well underway when the September 11 attacks occurred; the Amerithrax events and the public health crisis that followed gave added urgency to the Center's efforts.

One of the most important powers needed in catastrophic health crises concerns coercive powers: mandatory medical examinations and treatment of at-risk individuals, for example, or compulsory isolation or quarantining of infected persons, or even confiscation of contaminated property or human remains. Individual states have approached these questions in a continuing ad hoc fashion, with little regard for public health law in surrounding states. Jurisdictional disputes can usually be resolved with time, but in a bioterrorist attack, or a potential pandemic involving agents like the SARS or H5N1 avian flu viruses, time is the one thing we certainly will not have. The paralysis that can set in when local, state, and federal jurisdictions are unclear was glaringly apparent in the fumbled responses to Hurricane Katrina in 2005. The present tribalistic nature of individual state public health systems and laws could be a major impediment to rapid, coordinated, and effective federal-state response to a catastrophic health emergency. It could cost human lives.

The result of the work done by the Center for Law and the Public's Health is called the Model State Emergency Health Powers Act. It was crafted by the two faculties involved, but with extensive input from a workshop of public health, emergency services, and national security experts. A draft proposal coming out of these efforts was then circulated nationwide to an even broader range of public health and emergency response experts for additional review and comment before submission to the CDC.

It was called a *Model* Act because it was intended only as a guideline for states to consider as they crystallize their own thinking about what powers they might have to assume in a catastrophic health crisis along the lines of a Dark Winter. Subsequently considered and debated in nearly every state in the Union, the Act provoked considerable debate and even some opposition, revealing fundamental splits in the way many Americans (and their legislators) view the role of government in anything other than building roads. As in so many other American debates, the schism lay between collective action in the face of emergencies and protection of individual privacy and property rights; between concentration of power in the hands of an executive and individual choice in decision making.

The Act is intended to prompt states to develop comprehensive and integrable plans to deal with catastrophic health emergencies, and to define clearly the legal bases for actions by state and local agencies charged with handling such crises (Box 8.1).

BOX 8.1
KEY PROVISIONS OF THE MODEL ACT

Article I   Purposes and definitions.

Article II   Planning for a public health emergency. Calls
                  upon each governor to appoint a Public Health
                  Emergency Planning Commission; requires
                  Commission to present Governor with a plan for
                  responding to public health emergencies.

Article III   Detecting and tracking health emergencies.
                  Requires immediate reporting by doctors,
                  pharmacists, and veterinarians of unusual health
                  occurrences that might suggest a health emergency.

Article IV   Declaring a state public health emergency.
                  Empowers Governor of each state to declare a public
                  health emergency. Grants the Governor emergency
                  powers to suspend certain regulatory laws; to shift
                  responsibilities and personnel within state agencies;
                  to mobilize the National Guard; to seek aid from
                  other states and from federal government.

Article V   Special powers: property. Would grant governors
                  broad powers to seize, condemn, or order the
                  decontamination of property; to seize materials
                  and property required to meet the emergency; to
                  place health care facilities under control of state's
                  public health agency; to control disposal of human
                  remains; to institute and control rationing of
                  health-related products.

Article VI   Protection of persons. Grants governors the power
                  to compel examination, testing, and treatment of
                  affected individuals, and quarantining or isolation
                  of individuals and specific geographic areas.

Article VII   Communication. Requires each state's public
                  health authorities to inform the people of the state,
                  on a regular basis and by all available means, about
                  the declaration of an emergency, its expected or
                  estimated duration, steps being taken to control
                  the emergency, steps individual citizens should take
                  to protect themselves and their families.

The vast majority of states acted swiftly and have now upgraded their own systems and collaborated with other states to develop effective mutual aid plans.

In all of its provisions, the Model Act bends over backwards to show sensitivity to religious or ethnic attitudes, as well as to personal and medical privacy concerns. The powers conferred under full implementation of this Act are intended to be triggered only in an extreme emergency, and to expire with the emergency. There is ample precedent in American law for the temporary suspension of civil liberties. But if we are to prepare ourselves adequately for a real Dark Winter or a 1918-like flu pandemic, there are a number of issues apart from training first responders and stockpiling vaccines that we need to consider.

SOCIAL AND ECONOMIC CONTINUITY IN A NATIONAL HEALTH CRISIS

Neither a bioterrorism attack nor a nationwide flu pandemic will degrade our nation's physical infrastructure, but both—and in particular a widespread and long-lasting pandemic—will place an uncommonly severe strain on our social and economic infrastructures. Large-scale absences of employees involved in health care services, in provision of utilities (gas, water, electric), and in public safety, transportation, and many other critical areas will strain our social structure to its very limits.

Maintaining continuity, and some sense of normality under these conditions, will require careful planning. The National Strategy for Pandemic Influenza recommends that public and private employers anticipate up to 40 percent of their employees being unable to function at various times throughout a pandemic crisis. Not all of these may be ill. Some may be tending to family members who are ill; some may simply refuse to work in places they think are rife with disease. We can expect similar problems, although probably of shorter duration, in the event of a bioterrorism attack.

Private and public enterprises must begin to plan now for continuation of essential functions in major health crises, for their own and for the public good. We may have to practice *social distancing*—basically, keeping people wherever possible out of sneezing distance from one another in work spaces, schools, churches, and other public gatherings. People may be required to wear face masks in public—free supplies of these should be stockpiled in appropriate places. Alternate worksites should be explored to avoid employee crowding; people should be encouraged whenever possible to work at home via computer. Businesses and government offices will need to establish conditions for delegating authority and responsibilities as employees at all levels are removed from the workplace due to illness or family crises.

Of particular concern will be the ability of public safety officials to respond adequately to civil disturbances and breakdowns in public order. As was laid out in the Dark Winter exercise, these might arise as health care facilities are overwhelmed or as people vie for limited supplies of vaccines or antivirals. Individuals may try to force themselves or family members into health care facilities ahead of others; they may also try forcibly to remove family members from health care facilities where they have been isolated or quarantined. Stress brings out the best in some people, but the worst in others. This will place a further strain on police, whose own ranks may have been thinned by illness.

Deciding who has access to scarce medical supplies and treatment (medical triage) will present a major dilemma. For either a bioterrorist attack or a natural pandemic, we may find ourselves in a situation where there are not enough drugs or vaccine to treat everyone who, under ideal conditions, might have access to these treatments. Particularly in the case of a pandemic, initial supplies of a true pandemic vaccine (as opposed to pre-pandemic vaccines; see p. 125) will likely be insufficient to treat everyone who wants it. In the case of influenza, we do have drugs for treating infected persons. But if transport of medical supplies becomes disrupted in the chaos of the pandemic itself, getting these drugs to where they are most needed may result in localized shortages. Decisions will have to be made in such situations about who gets the limited supplies available. How will these decisions be made, and who will make them?

The overriding principle is in theory simple: drugs and vaccines should be distributed in such a way as to minimize the number of people dying. But that does not necessarily mean giving them to sick people lined up at the clinic door on a first-come, or even most-sick, first-served basis. Here are some of the considerations decision makers will have to juggle.

- Health care workers (doctors, nurses, emergency medical technicians) are essential to keep hospitals and clinics operating. Each such individual may be instrumental in saving dozens, maybe hundreds, of lives. Protecting these individuals will greatly increase the number of people who can be saved overall.
- People charged with maintaining civic infrastructure— those who know how to keep water, gas, and electricity coming; those involved in public safety (police and fire); those who know how to keep a city operating on a day-to-day basis—all of these will be critical in assuring that the cities and communities in which we will have to wage our battle against lethal pathogens continue to function. The chaos and confusion that would ensue should we lose these people would certainly end up costing many more lives than necessary.
- Hard decisions will still have to be made among those presenting themselves at the clinic door. This is the most emotionally traumatic form of triage (from the French *trier*—to sort out), and it means deciding whether the individual presenting can recover without treatment, needs treatment and can likely be helped by it, or is too advanced in a disease to benefit from treatment. This may apply to things like ventilators or other scarce ICU equipment, as well as to vaccines and medicines.

Triage is not a perfect science, but there are some guidelines, especially for pandemic influenza. Flu vaccines are best if given before infection, and certainly will have no impact on disease beyond forty-eight hours after someone is infected—just about the time they begin displaying symptoms. Persons already showing signs of the flu may thus be turned away from vaccination stations. Antiviral drugs are a little less certain. They work best in the first two to three days of infection; if someone is showing signs of advanced disease, the drugs probably should not be wasted. They may be given a placebo to ease their anxiety and move them on through the line, but not a drug that could truly save someone else's life. The federal government, through NPSI, has committed to drawing up a set of recommendations to assist states and local governments in determining how scarce medical resources should be distributed. It is anticipated that states and communities may adopt these as authority for triage procedures. As of this date these guidelines have not yet been made public.

But even with federal guidelines, these are still terrible decisions for someone to have to make, knowing full well that often they must make them using on-the-spot intuition and impression rather than hard laboratory evaluation. People will resist decisions not to treat them, some violently. But in many situations these decisions will have to be made, and those forced by circumstance to make them must be protected and held immune from any liability. There is an entire field of legal inquiry that deals with shielding those charged with making these and other decisions in emergency situations.

A number of other unique legal issues will also likely emerge in a national health crisis. There will almost certainly be individuals who will try to profit from the crisis by selling, on the Internet or by other means, useless "medicines" or pieces of medical equipment claimed to be necessary to protect against whatever pathogen is threatening. If smallpox or Ebola were used in a bioterrorism attack, there are at present no drugs that will help. But many people will not know that. If vaccines or antimicrobial drugs exist but are unavailable, desperation may drive some people to find unorthodox sources to fill their perceived needs. Some of the remedies obtained in this fashion will doubtless be bogus, but packaged to look like bone fide pharmaceutical products. Aside from being worthless, they may lead those who buy and use them to believe they are protected from disease and to take risks that endanger themselves and others. False claims and worthless merchandise pass through the Internet every day. Law enforcement officials may or may not have time to deal with this problem, but it could very well cost lives.

The law will also be tested in a national pandemic, or in a bio-terrorist attack involving a contagious pathogen, in dealing with a highly contentious public safety issue: the forcible confinement of infected people—quarantine.

QUARANTINE

Quarantine goes back at least as far as the biblical Old Testament. Leviticus counsels us to isolate lepers and burn their garments. This advice was taken to heart during the fourteenth-century episodes of bubonic plague. The word *quarantine* itself derives from the Italian *quaranta giorni,* the forty-day period of isolation for ships, their merchandise, and often their crews and passengers after their arrival in European ports if they were coming from plague-ridden cities. In local towns and cities where plague had broken out through flea bites (see chapter 3), mandatory isolation of people in their homes or in secluded locations was common. In one form or another, and for an increasing number of contagious diseases, quarantine was a common practice well into the twentieth century.

In the United States, quarantining was originally established as a colony (later state) right, which the federal government tried at various times to preempt to deal with outbreaks of diseases such as yellow fever and cholera. Through a long series of legal maneuvers, states and their counties ended up as the primary agencies responsible for domestic quarantine. But the federal government still retains full authority to impose isolation and quarantine on both U.S. citizens and foreign nationals to prevent importation of contagious disease (Box 8.2) at international ports of entry located in individual states. The main agency in charge of federal quarantine is the CDC's Division of Global Migration and Quarantine. The U.S. Customs Service and the Coast Guard provide enforcement if required. Violation of federally imposed quarantine under these conditions is a criminal offense. The federal government also claims the right to impose quarantine within states in situations that threaten interstate commerce.

The use of quarantine as a public health measure gradually fell off as our understanding of infectious disease became more

BOX 8.2
CONTAGIOUS DISEASES SUBJECT TO
QUARANTINE AT U.S. PORTS OF ENTRY

Cholera	Plague	Tuberculosis[a]	Yellow fever
Diphtheria	Smallpox	Influenza[b]	VHF[c]

[a]infectious forms only
[b]pandemic forms ony, e.g., H5N1 avian flu
[c]viral hemorrhagic fever

refined. By the early twentieth century we knew that not all infectious diseases (diseases caused by a bacterium, fungus, parasite, or virus) are contagious (readily passed from one person to another.) The expanded development of vaccines, especially for children, generated populations largely immune to most infectious diseases, further reducing the need for quarantine.

Still, many people alive today remember a time when quarantine signs would go up on the doors of homes sheltering someone in the midst of a contagious infection. Today, on those occasions when such steps are necessary, we tend to use the term *isolation* for the segregation of contagiously infected individuals. Patients with active tuberculosis are routinely isolated in hospitals. This is not normally a controversial procedure. The term *quarantine* is generally reserved for the segregation of individuals or groups of individuals presumed to be or suspected of being contagiously infected, until found to be free of disease. Mandatory quarantine is an action that can provoke a range of civil rights sensitivities, and for which health officials are most concerned about legal precedent and authority, particularly if violation of quarantine could be treated as a criminal offense.

The United States came perilously close to flirting with quarantine in the early years of the AIDS crisis. Once HIV was identified as the cause of AIDS and it became clear that AIDS was a communicable infectious disease, some groups called for the wholesale roundup and physical quarantine of all HIV-infected individuals. Cuba is the only state that has physically quarantined HIV-infected individuals almost since the inception of the AIDS pandemic. But in Germany, a federal judge called for tattooing HIV-infected persons and placing them in quarantine. A noted American conservative commentator published an opinion piece in *The New York Times* advocating the same procedure here.

But in general, there has been little public support for such measures throughout the world. In the United States, twenty-five states did revise their public health laws to allow various forms of restraint for persons who knowingly continued to engage in behaviors that endangered others after they learned they were HIV-positive. Fewer than a dozen HIV-infected persons were actually placed into physical detention of any kind. More commonly, they were required to attend counseling sessions, as a result of which in most cases they voluntarily altered their behavior. Nineteen of these same states also enacted criminal statutes making continued HIV-related behavior endangering others a crime (reckless endangerment was a common charge in such cases).

The issue of quarantining came up again during the SARS pandemic. In China, Hong Kong, and even Canada, large numbers of persons were in fact placed under formal quarantine. In China, around 10,000 persons were quarantined, with the threat of possible execution for those breaking their quarantine before they were officially cleared. In the United States, President Bush signed an executive order adding SARS to the list of communicable diseases subject to federal quarantine at ports of entry. The few confirmed SARS cases in the United States were subject to normal hospital isolation. Eighty-five individuals suspected of being infected apparently all cooperated in some form of voluntary but monitored isolation. Given the restricted number of cases in this country, quarantining on a scale that would threaten civil liberties never became an issue. But if an H_5N_1-derived form of avian flu infectious for humans should begin to approach epidemic status here, quarantine could become a procedure public health officials want to use.

The heterogeneity of state laws affecting isolation and quarantine, and the ambiguity of the relationship between federal and state law in these matters, worries many public health experts. The imposition of quarantine could result in numerous civil rights actions which, in a prolonged emergency like a pandemic, could compromise an effective response. That is why Article VI of the Model Act specifically, and in some detail, addresses the issue of quarantine. It would grant to state public health authorities the legal right to impose quarantine, and make failure to comply with authorities a misdemeanor crime. The state may also determine who has direct personal access to quarantined individuals, when, and under what conditions. Unauthorized people entering a quarantined building or area could themselves be subject to quarantine. To the extent possible, individuals subject to quarantine would likely be confined to their own home—together with their family. But if the numbers get overwhelming, those quarantined could end up being housed together in special areas.

The issues of isolation and quarantine have come up again recently with the emergence of drug-resistant forms of *Mycobacterium tuberculosis,* the causative agent of TB. AIDS patients are particularly susceptible to TB, and as TB-infected AIDS patients are treated with ever more powerful doses of TB-fighting drugs, we are seeing the emergence, through mutation, of so-called multi-drug-resistant strains of *M. tuberculosis* (MDR-TB), and the even more deadly extreme drug-resistant strains (XDR-TB strains).

In South Africa, fifty-two of fifty-three HIV-positive patients infected with these new TB strains died over a three-month period in 2005. There is less experience with these strains in healthy individuals, but cases have arisen. In late 2006, a man with MDR-TB who consistently refused to follow instructions for containing his disease was forcibly incarcerated in an Arizona jail cell to protect the public.[4] In late May 2007, a man in the United States somehow became infected with what was initially thought to be an XDR-TB strain. He was allowed to leave the country, traveling to several other countries and then back to the United States before authorities caught up with him and placed him under formal federal quarantine,[5] the first time this had been invoked since 1963. *M. tuberculosis,* even the XDR strains, are slow growing, and he apparently had not reached the stage where he was infectious. Public health authorities are examining their procedures to see why he was not placed under immediate surveillance or quarantine in the United States before traveling, or forcibly incarcerated, upon recommendation of the CDC or the State Department, in any of the cities he passed through. This individual was subsequently found to have MDR-TB rather than XTR-TB.

Quarantined individuals are guaranteed certain rights by the Model Act. They have the right to petition an appropriate court for relief, and the quarantining authority is obligated to inform them of this right and to provide information on how to do so. Quarantine beyond a ten-day period would automatically require court review and written approval. Quarantined persons cannot, for obvious reasons, appear in person to argue their own cases, but must be represented by a proxy.

While quarantining may indeed be necessary to prevent spread of an epidemic, this practice does have some inherent dangers. Persons subject to quarantine, remember, do not themselves show any sign of disease. If they did, they would be immediately isolated. Quarantined persons are only considered *likely* to be infected. So there is a real risk, in fact a likelihood, that some portion of those quarantined in their homes or in groups do not actually harbor the infectious agent. As soon as those who are truly infected exhibit disease symptoms, they will be removed to an isolation unit. But in the meantime they may have spread the infectious agent to others with whom they were quarantined, but who were not in fact infected at the time of their quarantine.

Imagine yourself in a situation of group quarantine—forced into close proximity with other persons who appear to authorities to be, as you are presumed to be, at high risk of being infected. You may feel strongly that you are not. But authorities know that if they don't choke off this epidemic at its source, and quickly, it will become a pandemic, and thousands—maybe millions—will die. What would you do if you were someone in authority, expected to do whatever is necessary to control the epidemic for the public good? What would you do if you were quarantined and still in apparent good health, and for good reason desperate to remain that way? These questions may well be asked one day, and it behooves us all to think about them ahead of time.

COMPULSORY VACCINATION: PROBABLY NOT

Vaccination is widely viewed as one of the most cost-efficient and effective public health strategies. There is no question that vaccination has reduced deaths from infectious disease over the past hundred years, particularly among children and young people. Smallpox was completely eliminated as a naturally occurring human infectious disease through a worldwide immunization program. Infectious diseases that formerly laid waste to as many as 10 percent of children before the age of three are now a distant memory. Yet today, a growing number of Americans resist the practice of vaccination, particularly compulsory vaccination of children as a condition for entering school. Such feelings cross all categories of race, education, gender, and income; about one in eight Americans declare themselves opposed to compulsory vaccination. Similar proportions of the populations of most European countries have traditionally opposed compulsory vaccination.

Since 1980, all fifty U.S. states have laws compelling vaccination of children against various vaccine-preventable diseases before they can be admitted to school. The particular vaccine requirements vary from state to state—only polio, diphtheria, and measles immunizations are required in all fifty. Moreover, all fifty states allow exemptions for medical reasons—a genetically impaired immune system, for example, and status as a transplant recipient or chemotherapy patient are everywhere allowable exemptions. But forty-eight states grant exemptions for religious beliefs, and twenty allow exemptions for personal philosophical reasons. Philosophical opposition to vaccination has been a rapidly growing reason for requesting exemptions in recent years, fueled largely by a suspicion that vaccines may not be safe for even completely healthy children.

Concern about vaccine safety has been a longstanding unifying theme for opposition to vaccination. Consider the following excerpt from a book written in 1900:

> The vaccination practice, pushed to the front on all occasions by the medical profession, and through political connivance

made compulsory by the state, has not only become the chief menace and gravest danger to the health of the rising generation, but likewise the crowning outrage upon the personal liberty of the American citizen....Compulsory vaccination, poisoning the crimson currents of the human system with brute-extracted lymph under the infatuation that it would prevent small-pox, was one of the darkest blots that disfigured the last century. [6]

For almost any vaccine, if one goes far enough back in the record, instances of harm caused by a vaccine can be found. More recently, there has been a persistent concern expressed about the standard single-shot vaccine MMR (measles/mumps/rubella) routinely administered to schoolchildren in most states. A paper published in 1999 raised the possibility that this vaccine might be responsible for an increase in autism-spectrum disorders noted during the 1990s. This was a well-reasoned hypothesis, and it triggered an immediate and intense investigation. It also caused a large number of parents to insist that this vaccine not be administered to their children, and many school districts honored their request. After several years of very intense study, it was concluded from epidemiological and laboratory data that such an association does not exist. The original paper and its claims were withdrawn by the authors. Nevertheless, to this day an unfortunately large number of parents still refuse to allow their children to be exposed to this vaccine.

Overall, it is beyond question that the risks posed to a child by vaccination are thousands, if not millions, of times less the than potential harm that can come from an infectious disease tearing through an unvaccinated young body. But the other thread that runs—deeply—through the opposition of many individuals today can also be found in the statement above, made over a hundred years ago: compulsory vaccination is viewed as "the crowning outrage upon the personal liberty of the American citizen." In this view, the right to expose a child to enormous medical risk is part of the "personal liberty of the American citizen." The book containing these and other strong statements was the bible for a good half-century of a small but persistent element that fought compulsory vaccination in this country tooth and nail. Just punch "compulsory vaccination" into the World Wide Web, and read articles like "Give Me Liberty, Or They May Give Me Death."

Not only are *individuals* protected from a particular microbe by bolstering their immune systems; when *populations* are immunized, individuals—even those not immunized—benefit from the huge reduction in the frequency of that microbe in the vaccinated general population. Decisions of individual parents to withhold vaccination of their children pose a grave threat not only to their own children but to every child that comes into contact with their children. In either case, this is not a victimless decision; it affects all of society. That is the basis on which court after court has upheld compulsory vaccination laws.

Given that the diseases cause by bioterrorism or pandemic pathogens can have mortality rate ranging up to 60 to 80 percent, most people will likely be willing, even eager, to be vaccinated. But past experience suggests there will be some who for medical or religious or philosophical reasons will decline to be vaccinated. What are we to do in such cases? What if one in eight persons thought to be infected refuses vaccination? How much will that compromise attempts to contain an epidemic that could kill untold numbers of innocent people? The provisions of the Model Act, if adopted, would provide a legal basis for forced vaccinations, but it is hard to imagine someone being physically restrained while a medical worker forces a needle into his or her arm.

CHAPTER 9

THE POLITICS OF
BIOTERRORISM IN AMERICA

HOW DID WE ARRIVE AT OUR CURRENT NATIONAL POSTURE regarding bioterrorism, especially in the context of other challenges our nation faces in the twenty-first century? The fear of bioterrorism has driven our leaders to spend over $40,000,000,000 since 2001 to defend us against it, even amid persistent doubts about the magnitude of the threat it poses.

Although concerns about bioterrorism clearly predate 2001, it is also clear that part of what has driven the near hysteria about bioterrorism since that time is its conflation with a larger "war on terror." Declaring war on something is a time-honored way in American politics to raise an issue to a level of unquestionable urgency. In some cases, like Nixon's "War on Cancer," apotheosis of an idea to a warlike status can summon up the afterglow of America's performance in World War II and harness America's energy for a noble purpose. More often, unfortunately, it breeds anxiety and fear. The White House says it has identified over 80,000 potential terror targets in America but for security reasons cannot say what those targets are. Zbigniew Brzezinski has made the point that the war on terror has actually created a climate of fear in America.[1]

Another part of the terror of bioterrorism is that unlike other weapons of terror and mass destruction—bombs, chemicals, nuclear devices—bioterrorism is based on things we cannot see and few of us understand. We are reliant on scientific experts to explain them to us, and that adds yet another layer of uncertainty, both for the public and for political leaders. Science is not pure, and scientific experts themselves have differing points of view—political points of view—about bioterrorism, just as they have differing points of view about global warming, or stem cell research, or the beginnings of life.

One of the reasons the United States was willing to sign the Biological Weapons Convention in 1972 was that our military establishment (and those of most other nations) had, on a mostly practical basis, lost interest in the utility of bioweapons. They were difficult to use and to control and, based as they were on biological organisms, subject to the vagaries of the differing environments in which they might have to be used. The instances of use of biological weapons in warfare are scant indeed, and showed no particular advantage over conventional weapons. Still, pressure to consider these same bioweapons as potential tools in the hands of unskilled terrorists continued to build long after the military decided to pass on them. How did this happen?

The pathway from rejection of bioweapons by our military to their whole-hearted embrace by our politicians as putative weapons of choice for terrorists wanting to harm us is long and complex, and almost entirely political. A few of the major events that helped shape this transition in our national thinking are summarized in the paragraphs that follow.[2]

1985

One of the early high-level alarms about the potential of bioterrorism to cause major damage to an America ill-prepared to defend itself against a catastrophic health threat is sounded. It comes from a special committee of the prestigious U.S. National Academy of Sciences, chaired by the distinguished Nobel laureate and then President of Rockefeller University, Dr. Joshua Lederberg. The subsequent report from this committee cites terrorist use of even low-level biological weapons against American civilians as a significant threat.

Lederberg, a microbial geneticist with impeccable credentials, had been an early and forceful advocate of an international treaty banning biological weapons. He had participated in the key 1975 Asilomar meeting and was one of the first, and most persistent, to warn of the dangers of bioterrorism and the possibility that the new field of genetic engineering could produce biological weapons of unimaginable destructiveness. He also served as a consultant to numerous intelligence and defense agencies, and would be a particularly influential voice among those urging the United States to take the threat of bioterrorism more seriously. At the same time, he was among the first to raise the specter of emerging natural pathogens as a major health threat.

1992

Defection to the United States of Soviet bioweapons expert Kanatjan Alibekov (later known as Ken Alibek). Alibekov had worked in biological weapons programs for nearly two decades, and at the time of his defection was second in command of Biopreparat, the Soviet agency for bioweapons research. He paints a picture of a massive Soviet effort, with tens of thousands of scientists at dozens of laboratories, working to develop new biological weapons and improve existing ones—notwithstanding that the Soviet Union was a signatory to the 1972 Biological Weapons Convention. He says the Soviets assumed we were doing the same. In addition to work on all of the conventional pathogens that would soon show up on the lists drawn up by the CDC, he tells his debriefers that the Soviets were also working on developing genetically modified pathogens for weapons use.[3]

Although Alibekov's defection will not be known to the American public until 1998, the essence of what he has to say about the Soviet program quickly disseminates through the higher levels of the administration and Congress. It will have a profound influence on policy makers from that point on. Alibekov convinces numerous U.S. politicians and institutions to sponsor a variety of anti-bioterrorism research projects, with himself at the helm, although most of these projects—along with much of the "hard intel" he passed on to his debriefers—will later be discounted.[4]

1992, 1993

Public awareness of bioterrorism is elevated by two reports issued from the Congressional Office of Technology Assessment (OTA). These reports address terrorism in general, but find that biological weapons might well be more attractive to terrorists than even nuclear devices. Both reports declare biological weapons in the hands of terrorists not only possible but likely, and urge the United States to prepare its civilian population to withstand such attacks.[5] In a subsequent OTA report issued in 1993, the statement is floated that an airplane disseminating 100 kilograms (about 225 pounds) of anthrax spores over a major metropolitan area could cause between one million and three million deaths. Although almost immediately challenged by any number of bioweapons experts, this figure, quoted repeatedly over the next ten years, contributes substantially to a growing hysteria about the dangers of bioterrorism. Later that year, Dr. Lederberg urges the Department of Defense "to take the initiative, together with the Centers for Disease Control and Prevention, in formulating a comprehensive plan for civil defense against a biological warfare attack."[6]

1995

The first major push in Congress to address the threat of bioterrorism comes through the collaboration of two U.S. senators, Sam Nunn (D-GA)[7] and Richard Lugar (R-NB). They had previously worked together to design legislation enabling the United States to work with the Russian government to stem the flow of former Soviet weapons of mass destruction and technical expertise to states hostile to the United States. In late 1995, Nunn and Lugar open hearings that result in a report entitled *Global Proliferation of Weapons of Mass Destruction*. These hearings are stimulated largely by the recent Aum Shinrikyo episode, and focus on the possible use of WMD, including bioweapons, as agents of terror against Americans on American soil.

Most of those who testify about bioweapons at these hearings support Lederberg's point about the woefully inadequate preparation of the U.S. public health and emergency response systems to deal with the consequences of *any* kind of terrorism attack, but bioterrorism attacks in particular. The overwhelming tone of the hearings is that a bioterrorism attack on American soil is a foregone conclusion: not a matter of *if*, but *when*. The Aum Shinrikyo story and the antics of Larry Wayne Harris are put forward as harbingers of things to come, and will be major factors in increasingly loud calls for America to mobilize its ability to defend itself against biological attacks.

1995

William Patrick (among others) speaks at a high-level meeting called by the Clinton White House to discuss the implications of the Aum Shinrikyo affair. Patrick, who headed a major section at Fort Detrick when it was still researching weapons for biological warfare, has unassailable credentials as a bioweapons expert. He was also a major debriefer of Ken Alibek after his defection. Patrick makes a very persuasive case for the reality of the threat of bioterrorist attacks in the United States, presenting a hypothetical attack on the World Trade Centers using weaponized botulin toxin dispersed through the buildings' ventilation system. He will continue to sound warnings about the dangers of bioterrorism across the next decade.

1997

Public awareness of the perceived threat of bioterrorism grows considerably when in August of this year the *Journal of the American Medical Association* devotes most of an entire issue to the subject. Numerous op-ed pieces in the journal by luminaries such as Joshua Lederberg and Defense Secretary William Cohen suggest that bioterrorism on American soil almost certainly cannot be averted. Various articles in this issue of the *Journal* describe the history of biological warfare, provide brief descriptions of conventional agents used in bioweapons, advise physicians on the clinical recognition of bioterrorism victims, predict the psychosocial consequences of a bioterrorist attack, and explore a number of related topics.

This focus of a highly regarded American medical journal on the still somewhat esoteric topic of bioterrorism receives tremendous attention in media outlets in America and throughout the world. Several months later, on an ABC news program, Secretary Cohen holds up a five-pound bag of sugar and declares that if it contained anthrax spores and were dumped over Washington, D.C., half the inhabitants of that city would be killed. One imagines his choice of a target city is not lost on his political colleagues.

1996–98

Two books and a television program affect the thinking not only of the public but of politicians. *The Cobra Event* describes release of a genetically engineered form of the smallpox virus that attacks the nervous system.[8] Tom Clancy's *Executive Order* presents a bioterrorist attack with an aerosolized Ebola virus.[9] Both books describe in lurid detail not only the effects on humans of the pathogens involved, but the extensive social and civic breakdown that follows the attacks. President Clinton confessed to having been greatly affected by *The Cobra Event*.[10] An hour-long ABC *Primetime Live* program called "Germ Warfare: Weapons of Terror" parades a series of experts declaring a bioterrorist attack in the United States to be a near certainty.

1998

The Institute of Medicine, part of the U.S. National Academy of Sciences, releases a report generally agreeing with the notion that bioterrorism presents a major threat to the American people, and points out that a great deal needs to be done to prepare the American public health and emergency response systems to deal with it.[11] The report notes that although this would require a huge investment of money, the upgrades that would result were long overdue, and would benefit Americans even if a bioterrorism attack never occurred, especially in terms of being able to defend themselves against emerging natural pathogens. HIV, and the devastation it continues to cause in the United States and around the world, is cited as a prime example.

1999

The Johns Hopkins Center for Civilian Biodefense Studies is established. The most prestigious of several such groups established with federal support, this multidisciplinary group of faculty makes the strongest and most coherent case for increasing preparedness for a bioterrorism attack. Their numerous policy papers, and exercises such as Dark Winter and Atlantic Storm, provide a major resource for those urging increased spending for bioterror defense.

As the millennium approached, wrote Jeanne Guillemin, an international security expert at MIT,

> influential politicians and consulting experts broadcast apocalyptic visions of thousands, even hundreds of thousands of Americans dying from unnatural, intentional epidemics of anthrax, smallpox, or some newly devised disease, invisibly inflicted by barbarous foreigners.[12]

2001

And then, of course, came September 11, and the subsequent anthrax attacks with their implications for the potential of bioterrorism.

Having listened to the growing chorus of dire warnings by impeccably credentialed scientific and terrorism experts, of which the above is but a brief sampling; having only months before digested summary reports from the Dark Winter exercise; having been told by Alibek and our own intelligence agencies that former Soviet bioweapons scientists were almost certainly working in rogue states to produce tools for bioterrorists to use against defenseless American cities—what were American political leaders to do?

If they supported a massive effort to build a credible defense against bioterrorism, as so many were calling for, and a bioterrorism attack never materialized, the worst that could happen is they might be accused of having squandered taxpayer dollars. Another federal boondoggle for companies that profited from the resulting buildup; hardly news. As one writer put it, "we should not have to wait for the biological equivalent of Hiroshima to rally our defenses."[13] And besides, the politicians could always claim it was the defenses they had built that had discouraged terrorists from even trying.

But if they opposed spending billions of dollars on creating a bioterrorism defense, and an attack occurred, and thousands—perhaps tens of thousands—of people died, what then? Would anyone praise them as prudent defenders of fiscal responsibility? Not likely. They would probably go down in history as the most incompetent set of leaders this or any other country had ever produced. Criminally negligent; maybe, in the minds of the public and subsequent historians, just this side of war criminals.

What would *any of us* have done?

What the U.S. government did was invest over $40,000,000,000 after September 11 to upgrade America's ability to defend itself against a bioterror attack. To some extent, this action also improved our ability to defend ourselves against any major outbreak of disease, including a flu pandemic. But closer scrutiny has led some government groups and many outside experts to raise serious questions regarding the usefulness of our preparations to defend ourselves against bioterrorism,[14] and indeed the magnitude of the threat of bioterrorism itself.

Beginning as early as 1997, the Government Accountability Office (GAO), a nonpartisan agency created by Congress to analyze the cost and effectiveness of federally funded programs, issued a series of cautionary statements about the growing funding for anti-terror programs generally, arguing that spending priorities had not been established. They also expressed concern that sound risk-assessment studies had not been carried out, goals had not been defined, and there was little coordination of spending on the many programs that had been approved. The GAO also questioned the likelihood of chemical and biological weapons use by terrorists, given the complexity of their production and utilization.[15] Two years later the GAO warned that risk assessment studies still had not been carried out, particularly for expenditures to combat bioterrorism, the dangers of which it continued to question.[16]

Individual political and scientific experts also raised cautionary flags about the direction and magnitude of our efforts to defend against bioterrorism. Milton Leitenberg, trained as a scientist and an arms control expert, prepared a detailed analysis of the Aum Shinrikyo episode, in which he showed that far from representing the dawn of a new and dangerous world in which "anyone could do it" bioterrorism would be a major threat to humanity, it suggested something far different:

> Aum utterly and totally failed [to produce biological weapons], after no small expenditure of time and money.... The experience of Aum is...in marked contrast to the legions of statements by senior US government officials and other spokesmen claiming that the preparation of biological agents and weapons could be carried out in "kitchens," "bathrooms," "garages," "home breweries," and is a matter of relative ease and simplicity. [17]

This point was echoed by Amy Smithson, when she was Director of the Stimson Center's Chemical and Biological Weapons Nonproliferation Project:

> The subject of unconventional [biological] terrorism was tailor-made for hyperbole, and unfortunately much of what has been said has made it difficult to ascertain the gravity of the unconventional terrorist threat. Taken together, the technical realities, actual case histories, and statistical records of terrorist behavior with chemical and biological substances undercut the rhetoric considerably and point not to catastrophic terrorism but to small attacks where a few, not thousands, would be harmed.[18]

Brian Jenkins, a terrorism expert with the RAND Corporation, has repeatedly questioned the magnitude of the threat posed by bioterrorism. He points out that, with the possible exception of

Al-Qaeda, there has been no indication that terrorist groups have seriously considered the use of bioweapons as agents of terror. But even Al-Qaeda's interest in bioweapons for terror purposes is open to question. Consider the following statement by Ayman al-Zawahiri, left behind on a computer in Kabul, Afghanistan after Al-Qaeda members had fled in 2001:

> The enemy starting thinking about [biological and chemical] weapons before WWI. Despite their extreme danger, we only became aware of them when the enemy drew our attention to them by repeatedly expressing concerns that they can be produced simply with easily available materials.[19]

Terrorism expert Bruce Hoffman summed up the years of frantic governmental spending on bioterrorism after it took off in the late 1990s:

> [Bioterrorism] was where the funding was, and people were sticking their hands in the pot. It was the sexiest of all the terrorism threats and it was becoming a cash cow. So the threat of bioterrorism became a kind of self-fulfilling prophecy. It was archetypical Washington politics in the sense that you generate an issue and it takes on a life of its own.[20]

Many tried to remind us that the real biological threat to Americans is, and will continue to be, emerging pathogens— viruses like HIV, and possibly H5N1 avian influenza, or other pathogens that may cross into the human species in the future, and against which we have no innate immunity. We have known about the threat of an H5N1 pandemic since the late 1990s. Had we spent $40,000,000,000 directly on preparing ourselves to deal with the threat of emerging pathogens, rather than many of the quasi-military defense items associated with bioterrorism, we would probably be as well prepared for a bioterrorist attack, and certainly better prepared for a natural pandemic. But would we have done it—would we have spent all that time and energy and money—without the visceral response evoked by a more easily visualized attack by hypothetical bioterrorists, without the umbrella of a "war on terrorism"?

That is an intriguing question, to be sure. Perhaps it is simply that threats from pandemics, like those posed by earthquakes, tornadoes, and hurricanes, can be viewed as part of the natural order of things. At some level we expect to be hurt by them, even killed. It's part of living on a still turbulent planet. Bioterrorism, on the other hand, is an attack on us by other humans, using a natural pathogen or something derived from one as a weapon, for the sole premeditated purpose of trying to harm or kill us. That gets our blood running.

Fair enough. But our response to bioterrorism has not been to seek out and destroy potential bioterrorists and their weapons. Some of that is doubtless happening off the public's radar screen, but that is not what we have spent $40,000,000,000 on. We have spent that money to prepare ourselves to survive a bioweapons attack by vaguely defined (but usually bearded and turbaned) foreign terrorists. So what is our bottom line? Is it to defeat bio-terrorists, or is it to save lives? Our actions point clearly to the latter. Preferably both, of course. But then why in the world do we think we should just lie down and wait to die in a natural pandemic?

Perhaps it's that word *natural* again.

An increasing number of voices are telling us it is time to move toward a more mature view of bioterrorism, to tone down the rhetoric and see it for what it actually is: one of many difficult and potentially dangerous situations America—and the world—face in the decades ahead. No nation has infinite resources, and we must accept that we will never be able to make ourselves com-pletely safe from every threat we face. We will have to make ratio-nal assessments of those threats we can identify, and apportion our resources as intelligently and effectively as we can to deal with them. Our final chapter takes a small step in that direction.

CHAPTER 10

ASSESSING THE THREAT

MAKE NO MISTAKE—A BIOTERRORIST ATTACK *COULD* HAPPEN. But how likely is it that it *will* happen? No one can ever know for sure; that is why some level of preparation is certainly essential. On the other hand, a crippling pandemic caused by a variant of the H5N1 flu virus, or some emerging pathogen we haven't yet heard of, certainly *will* strike us at some point. The question we must ask ourselves—and our leaders—is, what are the most intelligent and effective ways we can prepare ourselves for each of these unpredictable threats, and how much of our limited resources should we invest in defending ourselves against each?

In this chapter, we try to assess the threat of bioterrorism in three ways. First, we ask what we have learned from the three episodes of bioterrorism or biocrime described in chapter 2 that may be of use in judging the likelihood of future bioterrorist attacks. Second, we look at the possible pathogens terrorists might use against us, and how well prepared we are to defend against them. Finally, we ask who could—who would—carry out a bioterrorist attack against us.

I. WHAT HAVE WE SEEN SO FAR, AND WHAT HAVE WE LEARNED?

WHAT WE HAVE SEEN

THE RAJNEESH CULT, OREGON, 1984

Although it is arguable whether the attacks mounted by the Rajneesh cult should be considered bioterrorism, they have enough of the elements of what we might expect in a real bioterrorist attack that they are worth at least a brief visit.

WHAT DID IT TAKE TO MOUNT THE ATTACK? The agent, in this case *Salmonella enterica typhimurium* bacteria, was distributed directly, by hand, onto salad bar foods by cult members. It appears to have been in the form of a simple suspension of bacteria in a broth, carried in a closed culture tube, and poured directly onto the food. This is certainly an efficient method of delivery, but has obvious limitations for terrorists looking for dramatic TV footage.

WHERE DID THE BIOLOGICAL AGENT(S) COME FROM? The original salmonella cultures were traced to a scientific supply house in Seattle, of the type that normally supplies such cultures to university and medical center research laboratories, but would, at the time, sell them to anyone who ordered them. The Rajneeshees also purchased additional dangerous pathogens from the American Type Culture Collection (ATCC), a large East Coast facility that supplies medical and scientific labs worldwide with all sorts of cells—human, animal, and bacterial, plus viruses—for research purposes. The pathogens acquired by the Rajneeshees included those causing typhoid fever, gonorrhea, dysentery, and the CDC A-list pathogen *Francisella tularensis*. As far as is known, no large-scale production of these agents was carried out by the cult.

Today it would be impossible for an ordinary person to order any of these agents commercially, and difficult to pilfer them from a lab. The 2002 Bioterrorism Act requires both sellers and buyers of Select List pathogens to have a license. Moreover, laboratories that are approved to purchase Select List pathogens for research are responsible for what happens with them, and are required to secure them and to keep strict records of additions, removals, expansion, and transfers to other licensed laboratories. Failure to account fully for every organism under a laboratory's control could result in suspension of that laboratory's license. While a determined individual might still find a way to get at a lab's stock of Select List pathogens, it is much, much harder today than it was in 1984, or even 2001.

WHO PREPARED THE AGENTS USED IN THE ATTACK? Growth of an ordinary bacterium like *S. enterica typhimurium* can be carried out by anyone with a college-level laboratory course in microbiology. At the Rajneesh compound, a nurse (who would have had such a course) was the director of the Rajneesh Medical Corporation, which consisted of a clinic, a pharmacy, and a state-licensed medical laboratory. The cover provided by this latter facility made it relatively easy to order and receive biological materials and equipment. The laboratory was staffed by at least one licensed medical technician, who would have been specifically trained to grow bacteria.

One of the most important things we learned from the attempts of the Rajneesh cult to use bioterror to influence local politics was that, on the limited scale on which these individuals were working, with a pathogen requiring little skill to prepare for the uses they intended, they were able to cause substantial social and economic disruption in a small demographic area for a short period of time.

We also see in miniature in this episode what can happen when public health services are overwhelmed in an acute emergency. The only hospital in The Dalles, Mid-Columbia Medical Center, had just 105 beds at the time, with only about a dozen empty. A total of 45 people required hospitalization over a roughly ten-day period. Health care resources during this time were severely strained, reducing the level of care available to others who required medical attention. Today, under the same constant pressure to streamline health care experienced all across America, *Mid-Columbia has half the beds it had in 1984.*

AUM SHINRIKYO, 1995

Attempts to use biological terror agents for the furtherance of religious or political goals were taken to a new level by the Aum Shinrikyo cult. They mounted what is still described as the largest, most expensive effort ever seen by a non-state entity to develop biological weapons for the express purpose of mass murder.

The two pathogens Aum Shinrikyo tried most seriously to develop into lethal bioweapons were anthrax spores and botulinum toxin. Both are many magnitudes more deadly than the salmonella used by the Rajneeshees; botulinum toxin is the most poisonous biological substance known. Anthrax spores are considered by the CDC to be at the top of the list of weaponizable bacteria.

Aum Shinrikyo carried out at least seven bioterror attacks, four with anthrax and three with botulinum toxin. In each case, by their own subsequent admission, the goal was mass murder—as many casualties as possible. Had they succeeded in developing true weapons-grade pathogens and workable delivery systems, cult members could have killed thousands, perhaps tens of thousands, of people—certainly many more than died in the sarin gas attacks. A close study of their failure to achieve their aims may have a great deal to tell us about the shape of bioterrorism in the future.

WHAT WAS REQUIRED TO MOUNT THE ATTACKS? To prepare and execute successfully the kinds of bioterrorist attacks Aum Shinrikyo was aiming for requires solving an equation in three variables: state-of-the-art laboratory facilities; enough money for equipment and supplies; and people with the knowledge, experience, and technical skills to carry out the program.

The cult had one or more modern, well-equipped scientific research laboratories. These were well stocked with supplies used in microbiology research. Whether they were equipped and stocked properly for production of bioweapons cannot be ascertained from the descriptions we have. Clearly they had more than enough money available to undertake bioweapons production. But someone would have had to know precisely what is required to set up such a laboratory. That is where the people component of the above equation comes in. It is true that compared with the Rajneeshees, Aum Shinrikyo had an impressive staff of scientists in the inner circle. But were they up to the task? With all of their impressive credentials, what kind of preparation did they really have to develop and deploy weapons-grade anthrax and botulinum toxin?

The backgrounds of many of the leaders of Aum Shinrikyo are only sketchily known. Seiichi Endo was the cult's Minister of Health and Welfare, and played the lead role in trying to develop biological weapons. He had a doctorate in molecular biology and apparently did research in virology at Kyoto Medical School, one of the better biomedical research centers in Japan. He would have had some experience, perhaps considerable experience, in handling microorganisms. As a trained research scientist, he would also have had the ability to read the scientific literature and in principle extract from it everything needed to work successfully with both anthrax and botulinum toxin, including all of the problems one normally encounters in this kind of work.

WHY DID THEY FAIL? Aum Shinrikyo had money, laboratories, and a number of highly trained scientists. So why did they fail, after several years of trying, to produce a functional bioweapon?

Aum Shinrikyo's foray into bioweapons began with botulinum toxin. Seiichi Endo collected *C. botulinum* from soil on the Japanese island of Hokkaido and used this as starting material for the isolation and purification of botulinum toxin. This is one of two ways to acquire deadly biological agents for use as a weapon; the other is to steal them from a university or government laboratory carrying out research with these organisms.

University and government laboratories that undertake isolation of pure materials from microorganisms harvested from nature know that it requires a full range of skilled scientists who are experienced in each of the many phases required for isolation and purification of microbial products. For example, in the case of botulinum toxin, it is not enough to know that it is made by a bacterium called *C. botulinum* that lives in soil. As we have seen, there is not just one *C. botulinum;* there are at least seven different strains of this bacterium, each producing a different neurotoxin. Only three of these are toxic in humans. There are even regional substrains of these strains, each differing from the others in tiny but possibly important ways. It takes a real expert to properly identify the right germ to begin working with.

A specialized microbiology research laboratory will also have skilled protein biochemists, who can take a crude extract of bacterial toxin and put it through the many rigorously controlled steps needed to bring the toxin to a purity hopefully approaching 100 percent. Preparations of toxin contaminated with left-over bacterial cell-wall products, for example, may trigger a violent immune response which, while not directed specifically at the toxin, may in fact promote a generalized elevated state of nonspecific immunity in affected individuals that could destroy the toxin. This is a persistent (but well understood and controlled) problem in vaccine production.

It is also necessary to know something about the stability of the toxin in various degrees of purity so that it can be stored without losing potency. In the case of botulinum toxin, highly purified preparations are actually less stable than cruder preparations. Sometimes chemical preservatives may be added, but attaining just the right formulation of preservatives is tricky, and the formulas are among the best-guarded secrets of most bioweapons programs. Various methods of storage must also be tested, and the exact conditions for maintaining stability over time must be rigorously observed. But even under the best conditions, at ambient temperatures and humidity most biological preparations cannot be maintained beyond a month or two.

Then there is the problem of choosing a method of delivery that assures widespread dissemination of the toxin without damaging it. Here it is crucial to consult with a biotechnologist, a combination of biologist and engineer. Should the material be spread from a low-flying aircraft? Wafted by fan from the back of a moving truck or the roof of a building? Pushed through a high-pressure spray nozzle? This requires knowledge of both the material being disseminated and the mechanics of various disseminating devices. The effect of dissemination on potency of the toxin must be precisely determined.

And finally, one needs a knowledgeable meteorologist as part of the team. Wind strength and direction, temperature, moisture, and UV radiation from the sun, as well as many components of automobile and industrial air pollution, can have a profound effect on botulinum toxin. Wind must be considered not only in terms of lateral direction but in terms of temperature-driven updrafts as well. UV radiation from the sun can greatly reduce the viability of many biological reagents. Failure to take all of these variables carefully into account can decrease manyfold the effectiveness of aerosolized bioweapons.

Endo also encountered serious problems producing weaponized anthrax. It appears that in this case, he chose the second method for acquiring dangerous pathogens: he probably stole them. Investigations by the Tokyo police suggested that Endo had most likely obtained his anthrax seed cultures through a lower-level cult member with ties to a local university.

Getting anthrax bacteria (however acquired) to grow in the laboratory requires considerable knowledge of the requirements of different strains. This usually comes from years of experience, judging when cultures must be subdivided and recognizing from the appearance of the cells themselves what more may be needed— or perhaps what is present in excess—for optimum growth. These are not things necessarily picked up in a university-level microbiology lab course.

Even more skill is required to convert expanding cultures to conditions that encourage a high degree of spore formation without killing off huge amounts of bacteria. The conditions that trigger spore formation are often lethal for the bacteria if not managed properly. But even under the best of conditions, there will be more bacteria (living or dead) than good spores, and both living and dead bacterial cells must be gotten rid of in order to end up with the highest possible degree of pure spores that will germinate upon settling into human lungs.

Conversion of even high-grade anthrax spores to a form that can be efficiently disseminated is another very tricky task. Ideally, the spores should be released as a cloud of very dry individual spore particles that stay aloft in the air as long as possible. This is an important finishing step, which can be accomplished in a variety of ways, including coating the spores with electrostatic agents that cause them to repel one another and not form clumps. Single spores and small clumps of a few spores will stay aloft for some time upon dispersal, increasing the possibility of inhalation, but larger clumps will settle too quickly to the ground or other surfaces. Moreover, larger spore clumps, even if inhaled, tend to be less infectious because they get trapped in the hair and cilia that line the human airway and are destroyed.

So first of all the spores must be dried, which if not done properly can easily damage them. And once dried they must be stirred gently to break up any large aggregates, which can lead to loss of even more spores. Finally, truly weaponized spores would have to be coated with an electrostatic agent. But Endo chose instead to spray a liquid slurry of what he hoped was a lethal batch of anthrax spores from the roof of the cult's eight-story laboratory in the Kameido region of Tokyo. This was almost certainly doomed to failure. As Jonathan Tucker has pointed out,

> The capability to disperse microbes and toxins over a wide area as an inhalable aerosol...requires a delivery system whose development would outstrip the technical capabilities of all but the most sophisticated terrorists. Not only is the dissemination process for biological agents inherently complex, requiring specialized equipment and expertise, but effective dispersal is easily disrupted by environmental and meteorological conditions.[1]

A sample of the material sprayed that day was collected by police, and was finally analyzed at an American university in 1999.[2] It was found to contain a low level of anthrax spores and anthrax bacteria, plus many other types of bacteria that may or may not have been part of the material sprayed from the cult's building. A DNA analysis of the anthrax material showed that it was from the Sterne 34F2 strain, used to vaccinate farm animals against anthrax. It is essentially completely harmless in humans.

But even had Endo managed to steal a highly lethal strain of anthrax, given the form of his preparation (liquid slurry), its poor quality, the means used to disseminate it, and the lack of attention given to weather conditions, the amount of damage done to humans even in the immediate vicinity of the building would have been minimal at best. Delivery of weaponized biological agents presented a formidable challenge that remains a serious barrier to the use of biological weapons to this day. As even William Patrick, a staunch advocate of a vigorous bioterrorism defense program, admits,

> A dry [anthrax] product with the desired properties requires serious development with skilled personnel and sophisticated equipment... [While] Iraq successfully produced high quality liquids of anthrax and botulinum A toxin in quantity, their efforts to weaponize their agents were crude and far from successful....By analogy, if a dedicated nation such as Iraq had problems with agent delivery and dissemination, it follows that terrorists would also experience these problems, and at a higher level of intensity.[3]

So the reason for failure of Aum Shinrikyo's failure can be summed up in three words: people, people, people. It wasn't a lack of facilities. It wasn't a lack of money. It was a lack of people with years of experience in a dozen different scientific and engineering specializations, either as active participants or as consultants. As Amy Smithson, a chemical and biological weapons expert now with the Center for Nonproliferation Studies, has written:

> True, almost any scientist can produce a toxic chemical or grow a biological agent in a laboratory beaker, but the scientists most likely to overcome the demands of causing mass casualties are the particularly innovative and dedicated ones, the types who excel in the creative environments of industry and academia. That modern-day Thomas Edisons and Madame Curies would flock to the next Shoko Asahara or Timothy McVeigh begs skepticism.[4]

While there were a few among the Aum leadership with advanced training in various scientific and medical fields, none were experienced in microbiology or other biological sciences directly impinging on bioweapons development. None had the biotechnology background for large-scale pathogen production. None had the engineering skills to produce an efficient weapons delivery system.

Some in the U.S. government saw the evidence of Aum's dabbling in production of bioweapons as trumpeting a dangerous new escalation in the global threat of bioterrorism. Others saw it as evidence that producing effective bioweapons was not trivial, and likely beyond the capabilities of even the most technically sophisticated terrorists. The latter view did not prevail.

AMERITHRAX, 2001

The anthrax attacks of September through early November of 2001 are very different from anything we have seen before in the short history of modern bioterrorism, here or anywhere else. The scale of the Amerithrax attacks was modest, in some ways even by comparison with the Rajneesh caper. But it was more deadly; five people died, and a dozen or so more were seriously injured.

WHAT DID IT TAKE TO DO IT? The quality of the anthrax spore preparations found in at least some of the letters tells us this was not done along the lines of the Minnesota Patriots Council and their homemade ricin. They were not made by an amateur microbiologist or a survivalist chemist.

The Amerithrax terrorist(s) possessed a lethal form of anthrax—a purified spore preparation, completely dry and finely dispersed into individual spores or very small clumps of spores prior to dissemination, although they appear not to have been treated with electrostatic agents.[5] The anthrax spores sent to Senators Daschle and Leahy were, if not of the very highest level of purity and potency, certainly good enough to cause very serious problems. They were likely made in a government or possibly university research facility highly specialized in producing anthrax spores.

The choice of delivery system for the Amerithrax incident harkens back to the Rajneeshees. It bypasses all of the complications of choosing a dispersal device, and the damage such a device might cause. It obviates all of the hard-to-assess vagaries of the weather. It involves delivering the final weapon by hand; even better, by the hands of others—unsuspecting postal workers. As with the Rajneeshees' salmonella, the efficiency of this method is high: each of the recovered envelopes that could be tested had less than one gram of spores—enough to kill dozens of people, under the right conditions, and maybe more. It would be very difficult to achieve anywhere near that efficiency by wafting the spores through the air.

In the end, the most unsettling aspect of Amerithrax is that it was likely not carried out by a foreign terrorist. At present, all evidence suggests it was most likely an American scientist, who was either working or had worked in an American laboratory— possibly even a government laboratory—who carried out these attacks. The failure of the FBI to identify the individual involved suggests this was a highly sophisticated person who knew exactly what it would take to cover his or her tracks.

WHAT HAVE WE LEARNED SO FAR?

Those who think deeply about America's response to bioterrorism should be very clear about one thing. It is almost inconceivable that any terrorist organization we know of in the world today, foreign or domestic, could on their own develop, from scratch, a bioweapon capable of causing mass casualties on American soil. It just isn't going to happen. The isolation and purification of the requisite pathogens from nature, although theoretically possible, is far beyond the ability of all but a handful of advanced scientific laboratories in the world. So is development of a mass-scale delivery system for most pathogens.

It is still conceivable that terrorists could somehow manage to steal a high-grade pathogen and use a low-tech delivery system, as in Amerithrax. They could infect themselves, or coat their clothing, with unweaponized contagious agents like smallpox or Ebola and walk around in a crowded shopping mall sneezing on or rubbing up against unsuspecting citizens. Or terrorists might obtain a completed bioweapon, ready for use, from a third party. We will discuss that possibility shortly. But first, let's have a look at the pathogens terrorists might try to use against us, however obtained.

II. AGENTS OF TERROR: ARE WE READY TO DEFEND OURSELVES?

The above assessments of pathogen production or procurement and weaponization by terrorists suggest it will be extremely difficult to prepare and deploy an effective bioweapon. But admittedly, "almost inconceivable" is not the same as "absolutely impossible." What if someone succeeded? How well have the preparations described in chapter 7 prepared us to survive a bioweapons attack? Let's begin with the CDC A list.

SMALLPOX

Smallpox comes closest to posing a threat that could equal a 1918-like flu pandemic, or the one we fear could develop from the H5N1 avian virus. The Dark Winter exercise described in chapter 1, among other things, convinced political leaders that the United States needed to stockpile smallpox vaccine. Since 2001, about 100,000,000 additional doses of smallpox vaccine, based on the strains of vaccinia virus used up through the 1970s, have been produced and placed into the SNS. That is enough to vaccinate about one third of the U.S. population.

Nearly 200,000,000 additional doses of vaccine based on a less pathogenic form of vaccinia virus have also been produced, but have not yet passed the final hurdles in clinical testing. Nevertheless, the vaccine looks promising, so these doses have been packaged for use. The FDA would decide whether to deploy this vaccine in the event of a catastrophic smallpox attack on the United States. Given that the mortality rate for smallpox is about 30 percent, it seems likely most people would consent to accept the risk.

There is also a drug (SIGA 246) that strongly suppresses pox virus replication in animals. This drug had FDA approval for fast-track testing in humans, and Phase I clinical trials are currently underway. The addition of an antiviral drug to our armamentarium against smallpox will be an enormous step forward in defending against the possibility of a devastating terrorist attack with this agent.

It must be remembered that what made the Dark Winter scenario possible was the absence of smallpox vaccination in the U.S. population for the preceding thirty years, and only weak residual natural immunity. We cannot at present take smallpox off the list of bioterror threats but, with adequate supplies of an effective vaccine to contain outbreaks, we may very well, in the next year or two, be able to make it unattractive to terrorists as a bioweapon.

ANTHRAX

Since anthrax is caused by a bacterium, it can be treated by antibiotics. Cipro is the only antibiotic formally approved by the FDA for treating anthrax infections, but several other categories of antibiotics are known to be effective. Because anthrax infections are so rare in the United States, most currently available antibiotics have never undergone rigorous clinical trials for efficacy against anthrax. But there is now enough Cipro, along with other antibiotics, in the Strategic National Stockpile and on reserve through pharmaceutical vendors to quickly contain any bioterrorism incident involving anthrax.

Anthrax is not contagious, but we would like to have an effective anthrax vaccine to immunize persons in the vicinity of an anthrax attack. The only anthrax vaccine currently licensed for human use is called AVA. Unfortunately, it must be administered over a period of eighteen months to reach full effectiveness, and requires annual booster shots to maintain that effectiveness. Clearly this is less than ideal for managing a large-scale anthrax attack. Research on a better vaccine is proceeding with high priority in a large number of government and private laboratories around the world.[6] In the meantime, the United States, through Project BioShield, has placed AVA into the Strategic National Stockpile. A total of 10,000,000 doses have been deposited in SNS as of May 2006.

PLAGUE

Plague is also contagious, though less so than smallpox or the flu. Although at least bubonic plague can be treated effectively by antibiotics, containing its spread after a bioterrorist attack and preventing its evolution into a pandemic require that a vaccine be available. Moreover, recent outbreaks of plague in Madagascar and India, which were treated with antibiotics, indicate the emergence of drug-resistant strains of *Yersinia pestis,* the bacterium causing plague.

Since 2000, the United States, Great Britain, and Canada have been working cooperatively on development of a new plague vaccine, aided by Project BioShield. They have produced a candidate vaccine based on two recombinant proteins from *Y. pestis.*[5] This vaccine can completely block transmission of flea-borne plague in mice, even when *Y. pestis* is given intranasally to mimic aerosol exposure. Tests have also been carried out in monkeys, which tolerated the vaccine well and produced plague antibodies. Such a vaccine would likely not be used to treat infected individuals but to prevent spread of the disease in the larger population.

The vaccine has now entered clinical trials; the first batch of volunteers, in whom the vaccine is being tested for safety, were immunized in late 2005. The trials appear to be going well, but as we go to press, definitive results have yet to be reported out. If these vaccines are approved and enter the SNS, we can feel reasonably confident that plague will also become very unattractive as a bioterrorist weapon.

BOTULISM

At present, standard treatment for poisoning by botulin toxin consists of botulin antitoxin, a preparation of antibodies to the toxin produced in animals, usually horses. There is a plentiful supply of this antitoxin worldwide; a contract was recently awarded by the Department of Health and Human Services to a firm in Canada to place additional doses into the Strategic National Stockpile. Because botulin toxin poisoning is not contagious, there is less urgency for a vaccine. There is a vaccine currently licensed for use, but for a variety of reasons it has been used only to immunize researchers working on botulism and a limited number of military personnel who might face risk of exposure to botulin toxin. A new vaccine is currently undergoing clinical trials, and could be available for deposit in SNS in the next few years.

TULAREMIA

The only current vaccine for tularemia, called LVS, was obtained in 1956 from the Soviet Union, where it had been used successfully to immunize humans against the disease. Methods for producing the vaccine were further developed at the Salk Institute. Although tested in a few military personnel, where the U.S. version appeared to be effective with just a single injection, this vaccine has never received an FDA license for general use.

The National Institutes of Health has issued two new grants under Project BioShield totaling $60,000,000 to fund research into new vaccines for tularemia. But at best, a licensed new vaccine for tularemia is years away. Should there be a bioterrorist incident involving *F. tularensis* in the meantime, it would likely be managed with early intensive antibiotic therapy. The FDA might also decide to allow use of the existing LVS vaccine, perhaps in an intranasal aerosol form, as a supplementary treatment.

EBOLA AND MARBURG VACCINES

The Ebola and Marburg viruses are capable, under some conditions, of causing outbreaks of extremely deadly disease. However, although the underlying viruses have been around for over thirty years, we have not seen a major epidemic, let alone a pandemic. Even in the sometimes crude public health systems in which some of these outbreaks have occurred, the infections have been relatively easy to contain in the absence of a vaccine.

This gives us confidence that our public health systems should also be able to contain an outbreak should these viruses be used in a bioterrorist attack in the United States. Nevertheless, we will not feel completely comfortable until we have an effective vaccine that will limit spread of the disease.

Intensive research into possible vaccines for hemorrhagic fever viruses began in earnest in the mid-1990s, and in 2003 these efforts began to bear fruit. Several vaccines that provide excellent protection against both viruses in mice and monkeys have been produced using highly imaginative procedures for vaccine production.[6] At least one of these vaccines has been approved in the United States for clinical trials in humans. More trials will likely begin in the next year or two. Once a vaccine has been approved by the FDA and placed into the SNS, concern about the use of these two viruses as terrorist bioweapons will be greatly diminished.

The United States has made impressive progress in the past half dozen years in building a stockpile of drugs and vaccines that would greatly limit the damage terrorists might inflict using CDC A-list bioweapons. This doesn't mean we are completely immune to terrorist attacks with these weapons, but the knowledge that the damage done by such an attack would be considerably less than terrorists might have hoped for just a few years ago, together with the expense and tremendous difficulty involved in mounting an attack, could be a major deterrent for many terrorist groups.

WHAT ABOUT GENETICALLY ENGINEERED BIOWEAPONS?

We have not seen so far any genetically altered human pathogens that have approached the stage of weaponization. The former Soviet Union engaged in research into these kinds of weapons, but there is no data suggesting they ever actually produced a functional weapon. What would it take to achieve that?

As we said before, people—people with many different kinds of deep theoretical and technical expertise—are absolutely necessary to generate conventional bioweapons based on pathogens, such as anthrax, that have been studied for fifty years or more. This is triply true for genetically modified pathogens. Dissemination of statements like the following show a lack of understanding about what it would take to produce a high-quality, fully functional, deliverable biological weapon based on a genetically modified pathogen.

Today, anyone with a high-school education can use widely available protocols and pre-packaged kits to modify the sequence of a gene or replace genes within a microorganism; one can also purchase small, disposable, self-contained bioreactors for propagating viruses and microorganisms. Such advances continue to lower the barrier to biologic-weapons development.[7]

This is reminiscent of the statement by Senator Frist. In fact, anyone who has ever worked in a molecular biology research lab will tell you that the likelihood that an individual with little or no advanced training, or a group of terrorists, no matter how well funded or ideologically committed, would be able to design, assemble, and weaponize a genetically engineered pathogen has to be just about as close to zero as you can get.

As mentioned in chapter 4, the first researchers to reconstruct a pathogenic virus from scratch in a research laboratory actually had the subunits of the virus's genome synthesized for them by a commercial laboratory. They simply e-mailed the published sequence for the full genome to the company, indicating which subunits they needed to have synthesized. Some weeks later, the company mailed the finished subunits back to the lab, where they were assembled into functional viruses.

Could terrorists do the same thing? Possibly, although probably not in the United States. As with controls already in place for restricting access to standard pathogens, it is already next to impossible for someone not associated with a credentialed research laboratory to purchase this kind of service from the private sector. Biotech firms, who stand to profit from this kind of business, are themselves in the vanguard of developing such controls, and with federal help will soon have them in place. Terrorists could possibly find access to the needed materials in second-or third-world biotech companies, where such controls might not be in place. But having subunits of a viral genome in hand and having a complete, fully functional virus in hand are two different things. You don't just stir the fragments around, shake a few times, and— *voilà!*—out pops a virus. Orchestrating the correct alignment of the subunits in a finished assembly is a highly sophisticated process involving a great many very complicated steps.

And there is still the same problem that terrorists would have even with standard pathogens. Having in hand a highly purified, highly lethal biological pathogen of whatever origin, one of nature's own or one created de novo by humans in a modern research laboratory, is not the same as having a functional bioweapon. It is just the first step. As Ken Alibek, the former Soviet bioweapons expert, has said,

> The most virulent culture in a test tube is useless as a weapon until it has been put through a process that gives it stability and predictability. The manufacturing technique is, in a sense, the real weapon, and it is harder to develop than individual agents.[8]

As technology continues to improve, and more and more steps are automated, it is possible that someday relatively unskilled individuals will be able to genetically engineer a devastating pathogen, and turn it into a highly effective lethal weapon. In an infinite universe, anything is possible. Just don't hold your breath.

So, are we ready to defend ourselves? Box 10.1 suggests how the threat of bioterrorism today with various possible pathogens might compare with the threat level immediately after September 11. In compiling these estimates, we take into account improvements since 2001 in public health response to major health emergencies, pathogen surveillance and control, and other measures described in chapter 7, as well as recent acquisition of vaccines and drugs.

Such comparisons are obviously open to all sorts of discussion and disputation, but that shouldn't stop us (or anyone else

BOX 10.1

ASSESSING THE THREAT: CDC A-LIST PATHOGENS (BY DISEASE)

Pathogen	Threat level	Rationale
2001		
Smallpox	8	extremely contagious; very little vaccine; no drugs
Anthrax	5	dangerous, but difficult to produce; some antibiotics
Plague	6	moderately contagious
Botulism	4	not contagious; toxin difficult to produce
Tularemia	3	was never that much of a threat
Hemorrhagic fever	4	contagious, but no natural pandemic has emerged
Genetically modified pathogens	3	technology beyond terrorists
2008		
Smallpox	3	adequate vaccine; drugs in pipeline
Anthrax	2	greatly increased stocks of antibiotics
Plague	2	numerous vaccines in pipeline
Botulism	4	
Tularemia	3	should be manageable with antibiotics
Hemorrhagic fever	2	vaccines in pipeline
Genetically modified pathogens	2	technology beyond terrorists

who'd like to make a case) from trying. And note that none of the proposed current threat levels are zero; we aren't completely out of the woods yet for any of these pathogens. In an infinite universe...

III. WHO WOULD DO IT?

Although most experts consider a bioterrorist attack a low-probability event, not even the harshest critics of America's response to the threat of bioterrorism to date are willing to take the possibility of such an attack entirely off the table. So where, then, would such an attack come from?

INDIVIDUALS

When we think of individuals and bioterrorism, we are really talking about domestic terrorism. It is difficult to imagine an individual foreign terrorist producing and weaponizing a pathogen and bringing it to the United States to mount a deadly attack with no other help. But we are already reasonably certain that an individual per se is capable of mounting a bioterrorist attack (or at least of committing a biocrime) against us, because all evidence suggests the Amerithrax perpetrator acted alone.

The damage done in the Amerithrax attack, in the larger scheme, was limited, but it needn't have been. What if, instead of mailing anthrax spores to a handful of people, he or she had sent letters to 500 or 5,000 people, in ten different locations across the country? This is not beyond what a single determined individual could do. We might have seen a thousand dead instead of five, approaching the level of losses sustained in the World Trade Center and Pentagon attacks. The magnitude of the response to such an attack, mounted by the federal government and state and local responders, would perforce have been enormous, as would the resulting social and economic disruption. The fear and uncertainty generated among the public would have made any terrorist proud.

America has a history of lone avengers, individuals who believe they have a quasi-divine mandate to right some perceived wrong done in the world, usually to them. If some people die in the process, well, so be it. Collateral damage. In recent years, we have the examples of Theodore Kaczynski and Timothy McVeigh to remind us what determined loners can do in their quest for "justice." And it is almost impossible to stop the actions of single individuals. Most never committed a crime before, never came onto the radar screen of those looking out for the public's security. We may have to live with them as the price of a free and open society.

To the extent that creation and deployment of a bioweapon is an extraordinarily difficult task for even the most determined organized groups, we can feel reasonably assured that an individual, acting alone, is unlikely to be able to mount a major biological attack against America any time soon. But Amerithrax did happen, and that will rightly never be far from the minds of biosecurity analysts and our political leaders. Steps have been taken to assure that commercial access to pathogens of the quality used in that attack are placed beyond the reach of even the relatively few who know where to look for them in the first place. But as discussed in chapter 4, the recent renewed research into new and genetically modified pathogens for weapons defense, in addition to possibly violating the 1972 Biological Weapons Convention, also greatly increases the chance that these agents could accidentally—or purposely—make their way into the environment, and into the wrong hands.

And Amerithrax did happen. We will have to leave it at that.

GROUPS

When it comes to the possibility of bioterrorism by groups in America, there are groups and there are groups. Domestic groups like the Minnesota Patriots Council, the Aryan Nations,[9] the Identity Christians (an inspiration to Timothy McVeigh), and the innumerable so-called "militias" of overweight, middle-aged men that tramp through the woods on weekends all remind us that the disaffected do not always act alone. But like those who do, these groups generally believe they are acting to right some sort of wrong. As Jessica Stern, of the Council on Foreign Relations, has written:

The most likely [domestic terrorists] are religious and extreme right-wing groups and groups seeking revenge who view secular rulers and the law they uphold as illegitimate. They are unconstrained by fear of government or public backlash, since their actions are carried out to please God and themselves, not to impress a secular constituency.... [T]heir ultimate objective is to create so much fear and chaos that the government's legitimacy is destroyed.[10]

She might have added to her list some of the groups acting with a more leftist orientation. We have seen acts of violence committed by "ecoterrorists" in defense of nature, or opposed to perceived urban overdevelopment or gas-guzzling cars. There are those who are passionately opposed to nuclear power plants, genetically modified crops, or animal experimentation. It is always possible that some of the more extreme of these groups could resort to major acts of terrorism based on the use of bioweapons, although in the past such groups have generally refrained from using lethal force.

On the other hand, given the background of September 11, and to the extent most of these people are championing causes to which they hope to recruit large numbers of their fellow citizens, it would seem unlikely that groups of either the right or left would use any form of terrorism as a tool. They usually care about their cause, and such an act could destroy that cause in the public eye for decades to come. But who knows?

But when we speak of groups and the possibility of bioterrorism, of course the large pink elephant in the room is Al-Qaeda and its various cells, offshoots, and copycats. Much has been made of the fact that materials relating to biological weapons were recovered from Al-Qaeda training camps near Kandahar, Afghanistan, in December 2001. Milton Leitenberg has described these findings in detail in a recent analysis.[11] Among the items found were books on biological warfare and on microbiology, dating mostly from the 1950s and '60s. These would have provided some information relevant to bioterrorism, but that information would have been far from cutting-edge. There were also articles from scientific journals, some fairly recent at the time, on pathogens such as *B. anthracis*, *Y. pestis*, and *C. botulinum*, as well as hepatitis viruses. Among the papers found were clear indications that Al-Qaeda had recruited at least one PhD-level scientist to help them, although apparently mostly for procuring additional scientific information. There were no references to genetic engineering or recombinant DNA.

Also found were letters indicating Al-Qaeda may have sent someone to the U.K. with an unspecified amount of money for purchasing vaccines, perhaps for immunization of anticipated laboratory workers. There were also letters containing crude diagrams showing the general layout of a laboratory, a list of some equipment, and references to the need to train people to work in laboratory work. The writer was probably a Pakistani scientist; the intended recipient may have been an Egyptian. The writer says he had visited a laboratory in the U.K. where research on pathogens was carried out, and had attended scientific conferences on pathogens.

Another individual, identified as a Malaysian with a bachelor's degree in clinical laboratory science from a California university, appears to have been involved with developing plans for a bioweapons lab for Al-Qaeda in Afghanistan. He may have tried to obtain pathogen cultures. He was arrested in Malaysia on other charges, and there is no evidence he ever accomplished anything in the laboratory.

While all of this shows that Al-Qaeda was seriously investigating the possibility of building bioweapons, nothing among the recovered materials or any other subsequent intelligence gathered suggests it had ever gone beyond the planning stages, or that any pathogenic strains had been obtained or established in an Al-Qaeda–associated laboratory. There is certainly no indication they had put together a scientific team of the caliber necessary to assemble and deliver a bioweapon capable of inflicting mass casualties. They had not even come close to achieving what Aum Shinrikyo achieved, which was something less than impressive.

It is important to remember that single, mass-casualty attacks may not necessarily be the most effective, from a terrorist groups' point of view. Multiple smaller attacks, especially if spread out over time, and perhaps involving livestock and water as well as human targets, might be much more effective in promoting a sense of uncertainty, of helplessness and vulnerability, in a civilian population. The economic disruptions could be enormous. We still have to prepare ourselves for that.

In the end, what may well stop groups like Al-Qaeda from using bioweapons to achieve their aims against us is that it is just too much trouble. Not only are biological weapons exceedingly difficult to build and operate, the United States has now developed vaccines or drugs to counter most known conventional pathogens. Countermeasures for the rest should be available over the next few years. We have the Strategic National Stockpile, Push Packages, and vendor-managed inventories, as well as the ability to deliver these materials and more to an attack site within a matter of hours. We could suffer casualties, yes, but not mass casualties. Conventional bombs and chemicals are much easier to obtain and use, and can achieve much the same ends with less risk. Sophisticated terrorist groups may well agree with virtually all professional of the military establishments around the world that actually had effective bioweapons in hand: they are simply not worth the bother. For at least the near future, bioterrorism for Al-Qaeda and its ilk may be a non-starter.

STATES

What about so-called rogue states? Might they undertake the development of biological weapons, and use them themselves in covert operations against the United States or give them to terrorists to use (Table 10.1)?[12] Many universities in some of these countries have impressive levels of expertise in microbiology, molecular biology, and recombinant DNA technology. A number of their scientists were trained in the United States, Europe, Japan, or Korea, or even Cuba. There may exist, within some of these states, sufficient animosity toward the United States that pulling together the necessary experts and convincing them to attempt to develop genetically engineered bioweapons and appropriate delivery systems could be possible. This would take years, and huge amounts of money, for an uncertain outcome. It is not clear that even Iraq, which had an extensive, state-supported bioweapons program through the 1990s, had developed an effective delivery system for the most deadly conventional pathogenic

TABLE 10.1   Some Potential State Suppliers of Bioweapons

Country	Comments
China	Despite denials, suspected of having transferred bioweapons technology to Iran and other countries.
Cuba	Has sophisticated biotechnology industry, is suspected by some of having well-developed bioweapons program.
Egypt	Strong university microbiology programs. Allegations by Israel of bioweapons program.
Iran	Strong biotechnology base. Believed to be pursuing bioweapons program.
Kazakhstan	Home to many former Soviet bioweapons facilities. Status of these uncertain. Has never formally renounced bioweapons reserach.
Libya	Formerly had bioweapons program. Current status uncertain.
N. Korea	Presumed to have strong bioweapons program, but no reliable intelligence.
Pakistan	Strong biotechnology base. Status of bioweapons programs uncertain.
Russia	Strong background in bioweapons. Current status uncertain, particularly with respect to plague.
Syria	Good pharmaceutical infrastructure. Status of bioweapons programs uncertain.
Uzbekistan	Houses several former Soviet bioweapons facilities. Presumed to still hold stockpiles of many Select List pathogens.

agents, and no evidence at all that they had developed a weapon based on genetically engineered pathogens.

Would such states have the political will to underwrite such a program in the face of certain massive retaliation by the United States if they were discovered as the authors or facilitators of an attack against us? But perhaps more to the point, could political leaders in these states, which are rarely free and open democracies, live with the uncertainty that one day their terrorist proxies might very well turn these same weapons against them, in order to bring down what the terrorists consider a corrupt or "infidel" government? That would give almost all of these countries pause. The consensus among almost all terrorism experts is that the likelihood of rogue states placing high-grade bioweapons in the hands of terrorist groups is also just about zero.

So who would do it? To the extent that bioterrorism would ever be used against the United States, all bets are that it would come from organized terrorist groups such as, yes, Al-Qaeda. But it is very clear there is a huge, multifaceted barrier between us and a successful bioterrorist attack by any group. That barrier consists of the extreme difficulty of any such group building an effective bioweapon on their own and the improbability of any organized state providing them with such a weapon. It is conceivable that eventually, as some of the technologies involved are simplified, these difficulties could be overcome to some extent. But there is still the problem our own and other militaries had with state-of-the-art bioweapons provided to them ready-made by their own governments: they are difficult to use and control, and have less impact than chemical or nuclear weapons. So why bother?

We are also buffered by the impressive improvements we have made in order to absorb the impact of such an attack, should one happen. Although our public health system still has a way to go, it is in much better shape now than it was after 2001. And again, we now have stocks of vaccines and medicines that would greatly blunt the consequences of a bioterrorist attack. We would be better off with a few more vaccines, but we are close to having them. The certainty that even a large-scale bioterrorist attack would have the desired effect is much less now than it was ten years ago.

BIOTERRORISM IN CONTEXT

Bioterrorism is a threat in the twenty-first century, but it is by no means, as we have so often been told over the past decade, the greatest threat we face. The best way to look at bioterrorism is from a terrorist's point of view. If the aim is to kill as many people as possible, there are better—and, more importantly, simpler—ways to do it. If our leaders didn't draw the correct conclusions from Aum Shinrikyo, the terrorists probably have. After spending two years and millions of dollars trying to develop an effective bioweapon, the Aum people just gave up and used gas. Even that wasn't terribly effective, but it worked. If the aim is to cause social and economic disruption, bioterrorism could do the trick, but there is nothing to suggest that the level of disruption, the level of fear and uncertainty sowed, would be any greater as a result of an attack with bioweapons than with well-placed bombs, or even gas.

Alarmist rhetoric may have been necessary to push the issue of bioterrorism out into the open where it could be examined, and appropriate steps taken to meet the challenge. We have examined it, somewhat compulsively, and we have responded—rather excessively. Undoubtedly, in the process, there were those who saw an opportunity to enhance their personal wealth or power. But clearly most of those raising the alarm were motivated by genuine concern for our national well-being. It was a real issue; it is a real issue. And it is pointless to blame the politicians who overreacted. As we said in the last chapter, they may have had little choice.

But it is time to move on now to a more realistic view of bioterrorism, to tone down the rhetoric and see it for what it actually is: one of many difficult and potentially dangerous situations we—and the world—face in the decades ahead. And it is certainly time to examine closely just how wisely we are spending billions of dollars annually to prepare for a bioterrorist attack. No nation has infinite resources, and we must accept that we will not be able to make ourselves completely safe from every threat we face. So we will have to make rational assessments of those threats we can identify, and apportion our resources as intelligently and effectively as we can. Here are just two of the things beside bioterrorism we will need to consider.

EMERGING AND RE-EMERGING INFECTIOUS DISEASES

Lest we forget, the world as we speak is in the midst of one of the most serious natural pandemics since 1918: HIV/AIDS.[13] Roughly 15,000 per year still die of AIDS each year in this country (Figure 10.1). As 2005 drew to a close, the World Health Organization estimated that 40,000,000 people worldwide were HIV-infected or had full-blown AIDS. The vast majority of these

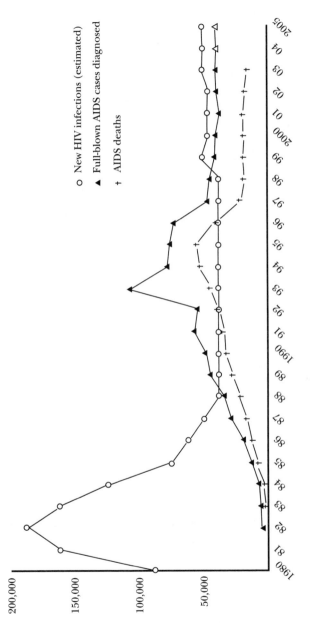

○ New HIV infections (estimated)

▲ Full-blown AIDS cases diagnosed

+ AIDS deaths

FIGURE 10.1 *The HIV/AIDS pandemic in the United States.*

will die prematurely of their disease. AIDS has already killed over 25,000,000 people worldwide. These numbers are beginning to approach those of the 1918 flu pandemic, and no cure is in sight. Two other infectious diseases, tuberculosis and malaria, account for 4,000,000 deaths annually worldwide. While malaria is essentially unknown in the United States, TB still kills a thousand people each year in this country. AIDS patients are particularly susceptible to TB, and as TB-infected AIDS patients are treated with ever more powerful doses of TB-fighting drugs, we are seeing the emergence, through mutation, of so-called multi-drug-resistant strains of *M. tuberculosis*, as well as the even more deadly extreme drug-resistant strains (XDR-TB strains). As recounted in chapter 8, XDR-TB is lethal in immunocompromised individuals, and could be spread to others once they become symptomatic. In late May of 2007, a man in the United States thought to be infected with an XDR-TB strain was somehow allowed to travel to half a dozen other countries before finally being quarantined upon his return to the United States.

And as we have seen, the world—including the United States—is now faced with the possibility of a form of avian flu (H5N1) that could, should it mutate or recombine with a common form of the human influenza virus in an individual simultaneously harboring both viruses, produce a virus that could equal the devastation wrought by the 1918 influenza virus—50,000,000 deaths or more worldwide before a pandemic played out. The impact of this on the American way of life we are so concerned about defending against terrorists is simply beyond the ability of any of us to comprehend. It is by far the most serious biological threat we face in the years ahead. The United States is finally mobilizing to meet this challenge; it has budgeted $7,000,000,000 to help prepare for the possibility of a worldwide H5N1-related pandemic. We can only hope that the same passion to act can be roused in our political leaders that was mustered for bioterrorism.

Admittedly, some of the money the United States has spent on preparing for bioterrorist attacks will enable us to respond more effectively to the threats posed by existing and emerging infectious diseases. It's time, however, to refocus our attention—and our resources and creative energies—more specifically toward some of nature's own threats, rather than depending on spin-offs from our concerns about bioterrorism.

GLOBAL CLIMATE CHANGE

It is now beyond dispute that the world, including the United States, is in for a period of global climate change whose extent cannot be predicted, and whose impact in terms of food supply, pathogen flow between species, and general social and economic disruption can only be guessed at. What is indisputable is that carbon dioxide levels have been rising steadily in the earth's atmosphere since at least 1960, and the surface temperature of the earth is rising in parallel. Glaciers and ice caps, which supply drinking and irrigation water to 40 percent of the earth's inhabitants, are disappearing at an accelerating pace.

It's essential that political arguments about what or who is causing global warming now take a back seat to discussions of how we are going to meet the challenges it forces on us. While assessments of bioterrorism, and to some extent even natural pandemics, are built on hypotheticals, global warming is *happening*, right now, and it will continue to happen for some time—maybe decades, maybe centuries. The social and economic disruptions accompanying a bioterrorist attack do not even show up as a single pixel on the screen of what will happen when the world's glaciers are gone and sea levels have risen twenty feet.

Will we see the same vigorousness (and even hysteria) brought to discussions of the increasingly severe hurricanes that will arrive along our southern and eastern seaboards in the years ahead? Have we really absorbed the lessons of Katrina? Do we really think the number of American lives lost in these kinds of events will be piddling in comparison to a bioterrorist attack? Will we see the same concerns raised about crop loss and economic disruption from increased temperatures and decreased water supplies that were raised about the possibilities of agroterrorism?

And what if we are struggling with the major social disruptions brought on by global warming and a major influenza pandemic at the same time? Could America as we know it survive?

None of this means that we should not continue to take prudent steps to defend ourselves against attacks with bioweapons.

We have lost five people so far to bioterrorism in this country. It is possible we could lose more if terrorists were to one day succeed in mounting a major attack using bioweapons. It has become a commonplace for most of us to say our world changed forever on September 11, 2001. Some politicians have told us the American way of life itself is in jeopardy; others, that September 11 was the first battle of World War III. This presumably is a war based on terrorism, and the terrorists, they tell us, are becoming ever more sophisticated.

The reality is that since September 11 terrorism of all types against the United States, on U.S. soil has not become easier; it has become a great deal harder. Part of the reason for that is the six to seven billion dollars per year we have spent over the past five years to upgrade our defenses. But how much longer should we continue to spend at that rate for biodefense? Has this become a permanently renewable check we must write each year? Let's hope the GAO continues, and even increases, its already critical scrutiny of biodefense spending, and forces Congress to take a stronger oversight stance.

In the meantime, we continue to lose fifty thousand people a year just to AIDS and seasonal flu. And we are facing challenges in the years ahead that make bioterrorism pale by comparison. So let us put bioterrorism in context. Even those who argued most strongly for a vigorous effort by the United States to defend itself against bioterrorism never considered bioterrorism more than a low-probability, high-consequence possibility. Realistic assessments of what is involved in making and using a bioweapon suggest the probability is even lower than many thought. Efforts made over the past decade to better defend ourselves against an attack, whatever its probability, have reduced the anticipated consequences.

It's time to move on to other things.

NOTES

CHAPTER 1

1. You can read the official final script of the Dark Winter exercise at http://www.upmc-biosecurity.org/pages/events/dark_winter. The information in this chapter was extracted (with minor dramatic license) from this script and from a published follow-up analysis of the exercise: Tara O'Connor, Michael Mair, and Thomas V. Inglesby, Shining Light on "Dark Winter," *Clinical Infectious Diseases*, 34(2002):972.

2. The transmission rate of ten for a primary smallpox outbreak, for example, is likely half or less of that depicted in Dark Winter: Raymond Gani and Steve Leach, "Transmission Potential of Smallpox in Contemporary Populations," *Nature* 414(2001):748. See also H. Pennington, "Smallpox and Bioterrorism," *Bulletin of the World Health Organization* (2003) 81:762. Also, improvements in home as well as hospital care developed while the disease was still prevalent seem to have been ignored. For a detailed critique of exercises such as Dark Winter, and in particular Atlantic Storm, see Milton Leitenberg, *Assessing the Biological Weapons and Bioterrorism Threat* (Carlisle Barracks, PA: U.S. Army War College, Strategic Studies Institute, 2005), 48–59. Available free at www.strategicstudiesinstitute.army. mil

CHAPTER 2

1. See for example Leitenberg, *Assessing.* We will discuss this issue further in chapter 10.

2. Brian Jenkins, "Will Terrorists Go Nuclear?" *Orbis* 29 (Autumn 1985):511.

3. Numerous accounts have been written about the Rajneesh cult, but the most accurate is probably that found in Seth Carus, "The

Rajneeshees," in *Toxic Terror: Assessing Terrorist Use of Chemical and Biological Weapons*, ed. Jonathan B. Tucker (Cambridge, Mass.: MIT Press, 2001).

4. We will look more closely at the biological agents used in these and other attacks, and additional agents considered by the U.S. government as being of particular concern, in chapter 3.

5. Claims that upwards of 5,000 were injured in these attacks are clearly exaggerated. Most of these excess claims were the so-called "worried well," who reported to hospitals to determine whether they might have been affected.

6. For a detailed history of Shoku Asahara and the Aum Shinrikyo cult, see David E. Kaplan and Andrew Marshall, *The Cult at the End of the World* (New York: Crown Publishers, 1996).

7. Like the Rajneeshees, Aum Shinrikyo established seemingly genuine businesses and medical facilities that made it easier to order potent chemicals and biologicals without raising suspicion.

8. This connection may have been suggested by the belief (later largely discounted) that the terrorists who bombed the World Trade Center in 1993 had included potassium cyanide in the explosive device they used.

9. For details of the Patriots Council and the Alexandria incidents, see Jonathan B. Tucker and Jason Pate, "The Minnesota Patriots Council," in *Toxic Terror*, ed. Jonathan B. Tucker (Cambridge, Mass.: MIT Press, 2001).

10. For more details on Harris, see Jessica Eve Stern, "Larry Wayne Harris, 1998" in Tucker, *Toxic Terror*.

CHAPTER 3

1. Countries alleged to have continued their programs include China, Egypt, India, Iran, Iraq, Libya, North Korea, Soviet Union/Russia, South Africa, South Korea, Syria, and Taiwan. Iraq, South Africa, and Russia may no longer have functional programs. Dispersal of former scientists working in the Russian/Soviet Union program to "rogue states" remains a major concern. For further details see Jeanne Guillemin, *Biological Weapons: From the Invention of State-sponsored Programs to Contemporary Bioterrorism* (New York: Columbia University Press, 2005); Ken Alibek, *Biohazard: The Chilling Story of the Largest Covert Biological Weapons Program in the World* (New York: Random House, 1999).

2. For more information on the pathogens discussed here, and the body's immune response to them, see William R. Clark, *In Defense*

of Self: How the Immune System Works in Health and Disease (New York: Oxford University Press, 2007).

3. Antibiotics effective against anthrax include ciprofloxacin (Cipro), tetracyclines such as doxycycline, and certain penicillins, like procaine penicillin G. To be effective, however, these drugs must be administered very early in the infection, especially in the case of inhalation anthrax. Rapid treatment of inhalation anthrax could probably reduce mortality even more.

4. Viruses do not usually receive a Latin genus–species designation, ordinarily reserved for bacteria and other cellular forms of life. However, the causative agent of smallpox was long presumed to be a bacterium, and the name given to the hypothetical agent, *V. major,* stuck after it was determined to be a virus.

5. The only viral diseases for which we have virus-specific drugs are the herpes viruses, the influenza virus, and HIV.

Chapter 4

1. A genome is the entirety of the DNA taken from a given organism, which in effect contains the full blueprint for the construction and operation of that organism. A recombinant genome is one that also contains one or more genes from a different organism.

2. Alibek, *Biohazard.* See also Janet R. Gilsdorf and Raymond A. Zilinskas, "New Considerations in Infectious Disease Outbreaks: The Threat of Genetically Modified Microbes," *Clinical Infectious Diseases* 40(2005):1160.

3. For a detailed discussion of what we know about the Soviet foray into genetically modified pathogens, see footnote 2 and Judith Miller, Stephan Engelberg, and William Broad, *Germs: Biological Weapons and America's Secret War* (New York: Simon and Schuster, 2001).

4. This has now been accomplished with the genome of *Mycoplasma genitalium.* See Karen Kaplan, "A Step Closer to Creating Life Out of Chemical Soup." *Los Angeles Times,* January 5, 2008.

5. For a fuller discussion of how the immune system is often the real culprit in disease, see Clark, *In Defense of Self.*

6. See http://dspace.mit.edu/bitstream/1721.1/32982/1/SB.v5.pdf

7. See for example http://www.etcgroup.org/article.asp?newsid=563

8. Hillel W. Cohen, Robert M. Gould, and Victor W. Sidel, "The Pitfalls of Bioterrorism Preparedness: The Anthrax and Smallpox Experiences," *American Journal of Public Health* 94(2004):1667.

9. Quoted in Leitenberg, *Assessing.*

CHAPTER 5

1. *Epidemic* is usually defined as a sudden, rampant spread of an infectious disease among humans within a single country, or possibly adjoining countries. When the disease involves a large number of countries or more than one continent, we use the term *pandemic*. The corresponding terms for movements of infectious diseases among animal populations are *epizootic* and *panzootic*.

2. Source: www.globalaging.org/health/us/fluaids.htm

3. In the first decades of the twentieth century, scientists did realize that some infectious agent existed that was much smaller in size than a bacterium, capable of passing through an extremely fine filter that trapped bacteria and incapable of being seen in a microscope, but they had no idea what it was. They referred to these agents as viruses. The influenza virus itself was discovered in 1933.

4. For an excellent and detailed account of the havoc wrought in the United States and elsewhere by the 1918 flu pandemic, see Gina Kolata, *The Story of the Great Influenza Pandemic of 1918* (New York: Simon & Schuster, 1999).

5. An influenza vaccine was developed shortly after World War II, and by the mid-1950s most doctors and public health officials were familiar with its use.

6. Although the earliest cases were reported from the relatively open Hong Kong, officials there insisted this flu originated in the more secretive People's Republic of China, probably in the adjacent province of Guangdong.

7. Over those that would be expected from normal seasonal flu.

8. In fact, for quite some time information about SARS in China was issued by the propaganda arm of the Chinese Communist Party rather than the government's Health Ministry.

9. For a detailed analysis of the pandemic as it played out in Hong Kong and Toronto, see C. David Naylor, Cyril Chantler, and Sian Griffiths, "Learning From SARS in Hong Kong and Toronto," *Journal of the American Medical Association* 291:2483–87.

10. David M. Bell, "Public Health Interventions and SARS Spread," *Emerging Infectious Diseases* 10(2004):1900–06.

11. Anyone who has recently been infected by an influenza virus will have antibodies in their blood specific for that particular virus, and these can be readily detected in a simple laboratory test.

12. In addition, it has now been shown that a pregnant H5N1-infected woman had transmitted the virus to her fetus. See J. Gu et al.,

"H5N1 Infection of the Respiratory Tract and Beyond: A Molecular Pathology Study." *Lancet* (2007) 370:1106.

13. For a Dark Winter–like description of what an H5N1-based pandemic could look like in a major metropolitan area like New York, see Irwin Redlener, *Americans At Risk* (New York: Knopf, 2006), 29–36.

14. See http://www.pandemicflu.gov/vaccine/prioritization.html#Draft GuidanceonAllocatingandTargetingPandemicInfluenzaVaccine

15. H. Markel et al., "Nonpharmaceutical Interventions Implemented by U.S. Cities During the 1918–1919 Influenza Pandemic," *Journal of the American Medical Association* (2007) 298:644.

16. Quoted in Scott Shane, "U.S. Germ-Research Policy Is Protested By 758 Scientists," *New York Times,* March 1, 2005.

17. Although not nearly enough to deal with a flu pandemic. There are roughly 100,000 ventilators in the United States; during a typical seasonal flu outbreak, the vast majority of these are in use. A major pandemic could easily require double the present number.

CHAPTER 6

1. The government actually carried out an agroterrorism exercise in two sessions in 2002 and 2003. This exercise, called Silent Prairie, was considerably scaled down from Dark Winter; there was no role-playing, for example. Whether because of perceived lack of public interest or for other reasons, few analytical details of this exercise have appeared in print. An even more abbreviated exercise called Crimson Sky has also not been made public. For Dark Summer I have drawn on what is known of these two exercises, as well as the very real British FMD outbreak of 2001. The excellent RAND Report by Peter Chalk, *Hitting America's Soft Underbelly: The Potential Threat of Deliberate Biological Attack Against the U.S. Agriculture and Food Industry* (MG-135-OSD, 2004), is also a valuable resource.

2. Vaccination was used in a few regions, but created complications when the epidemic was over in other regions because of the uncertain status of the vaccinated animals (see Dark Summer scenario above).

3. Most plant pathogens are harmless in humans, but there are some that can affect humans. For example, aflatoxin, made by a plant fungus, can cause liver damage and cancer in humans.

4. For an excellent and detailed analysis of potential terror threats to our water supply, see J. Nuzzo, "The Biological Threat to U.S. Water

Supplies: Toward a National Water Security Policy," *Biosecurity and Bioterrorism* 4:147 (2006).

CHAPTER 7

1. The complete Act can be viewed at http://www.fda.gov/oc/bioterrorism/bioact.html

2. For more information on the SNS, go to http://www.bt.cdc.gov/stockpile/. It is clear that concerns about bioterrorism (even before the Amerithrax incidents) were a major impetus in the creation of SNS. Initial funding for expanding SNS was provided through the Bioterrorism Act of 2002.

3. A detailed description of BioWatch is contained in Congressional Research Service Report RL 32152, available at www.fas.org/sgp/crs/terro/RL32152.html

4. Each of these laboratories is part of the Laboratory Response Network for Biological Terrorism, a network of a hundred or so laboratories established in 1999. Certified in advance by the CDC, they now cover every major metropolitan area.

5. For a complete analysis of state preparedness for catastrophic health emergencies, see http://healthyamericans.org/bioterror06/BioTerrorReport2006.pdf

6. For a detailed analysis of the most recent bioterrorism budget, see http://www.armscontrolcenter.org/resources/fy2008_bw_budget.pdf

7. The complete NSPI can be viewed online: www.whitehouse.gov/homeland/pandemic-influenza.html; www.pandemicflu.gov/plan/pdf/panflu20060313.pdf; www.whitehouse.gov/homeland/pandemic-influenza-implementation.html. An analysis of the early actions taken under NSPI can be found in Stephen S. Morse, "The U.S. Pandemic Influenza Implementation Plan at Six Months," *Nature Medicine* 13(2007):681–84.

8. The number of actual pills (as opposed to courses) required would depend on how the drug is used. Someone thought to be infected would take one course, or two pills per day for five days. Those wanting to protect themselves against possible infection would take consecutive courses until the possibility of infection recedes.

9. These are standard OSHA-approved respirators used by many craftspeople, and available to the public at outlets such as Home Depot for a very modest cost.

CHAPTER 8

1. Institute of Medicine, *The Future of Public Health* (Washington, D.C.: National Academy Press, 1988).

2. The complete Act can be viewed at http://www.publichealthlaw.net

3. Authorized under U.S. Code 264, Title 42.

4. Nicholas Riccardi, "The Man in the Leg Irons and Mask," *Los Angeles Times*, May 2, 2007.

5. John Schwartz, "Tangle of Conflicting Accounts in TB Patient's Odyssey," *New York Times*, June 2, 2007. This individual's TB was later determined to involve a somewhat less deadly, but still dangerous, MDR (multi-drug-resistant) strain of *M. tuberculosum*.

6. J. M. Peebles, *Vaccination a Curse and a Menace to Personal Liberty, With Statistics Showing Its Danger and Criminality* (Battle Creek, Mich.: Temple of Health Publishing, 1900).

CHAPTER 9

1. Zbigniew Brzezinski, "Terrorized by 'War on Terror'," *Washington Post*, March 25, 2007.

2. For an excellent account of the intense political battles during the 1980s and '90s that underlie much of the current official view of bioterrorism, and from which some of what follows was taken, see Susan Wright, "Terrorists and Biological Weapons: Forging the Linkage in the Clinton Administration," *Politics and the Life Sciences* 25(2007):57–115. See also Guillemin, *Biological Weapons*.

3. For a detailed account of Alibekov's defection and debriefing by U.S. intelligence experts, see also Judith Miller, Stephan Engelberg, and William Broad, *Germs: Biological Weapons and America's Secret War* (New York: Simon and Schuster, 2001).

4. David Willman, "Selling the Threat of Bioterrorism," *Los Angeles Times*, July 1, 2007.

5. U.S. Congress, Office of Technology Assessment, *Technology Against Terrorism: The Federal Effort* (Washington, D.C.: U.S. Government Printing Office, July 1991); U.S. Congress, Office of Technology Assessment, *Technology Against Terrorism: Structuring Security* (Washington, D.C.: U.S. Government Printing Office, January 1992).

6. Quoted in Wright, "Terrorists and Biological Weapons," 68–69.

7. Senator Nunn had played the role of the President in the Dark Winter exercise.

8. Richard Preston, *The Cobra Event* (New York: Ballantine, 1997).

9. Tom Clancy, *Executive Orders* (New York: Putnam, 1996).

10. Wright, "Terrorists and Biological Weapons," 83–84.

11. National Academy of Sciences, Institute of Medicine, *Improving Civilian Medical Response to Chemical or Biological Terrorist Incidents, Interim Report on Current Capabilities* (Washington, D.C.: National Academy of Sciences, 1998).

12. Guillemin, *Biological Weapons*, 149–50.

13. Steven M. Block, "The Growing Threat of Biological Weapons," *American Scientist* 89(2001):28–37.

14. For a critical analysis of some of the spending in the defense against bioterrorism program, see Redlener, *Americans at Risk*, chap. 7.

15. Government Accountability Office, *Combating Terrorism: Federal Agencies' Efforts to Implement National Policy and Strategy*, Report No. GAO/ NSIAD-97-254, September 1997.

16. Government Accountability Office, *Combating Terrorism: Need for Comprehensive Threat and Risk Assessments of Chemical and Biological Attacks*, Report No. GAO/NSIAD-99-163, September 1999.

17. Milton Leitenberg, "Aum Shinrikyo's Efforts to Produce Biological Weapons: A Case Study in the Serial Propagation of Misinformation," in *Terrorism and Political Violence* 11(1999):149–58.

18. Amy Smithson and Leslie-Anne Levy, *Ataxia: The Chemical and Biological Terrorism Threat and the US Response*, Stimson Center report no. 35, October 2000, 282.

19. Alan Cullison, "Inside Al-Qaeda's Hard Drive," *Atlantic Magazine*, September 2004.

20. Quoted in Wright, "Terrorists and Biological Weapons," 100.

CHAPTER 10

1. Jonathan B. Tucker and Amy Sands, "An Unlikely Threat," *Bulletin of the Atomic Scientists* 55(1999):46–52.

2. See Hiroshi Takahashi, et al., "*Bacillus anthracis* incident, Kameido, Tokyo, 1993," *Emerging Infectious Diseases* 10(2004):117–20.

3. William Patrick, "Biological Terrorism and Aerosol Dissemination," *Politics and the Life Sciences* 15(1996):208–10.

4. Smithson and Levy, *Ataxia*, 280.

5. See Gary Matsumoto, "Anthrax Powder: State of the Art?" *Science* 302(2003):1492–97.

6. For details of new strategies for vaccines against natural or man-made pathogens, see Clark, *In Defense of Self.*

7. David A. Relman, "Bioterrorism—Preparing to Fight the Next War," *New England Journal of Medicine* 354(2006):113–152.

8. Alibek, *Biohazard,* 97.

9. A visit to the Aryan Nations website is instructive. Their official motto now is "Violence Solves Everything," and they speak, ironically, of "Aryan Jihad."

10. Jessica Stern, "The Prospect of Domestic Bioterrorism," *Emerging Infectious Diseases* 5(1999):517–22.

11. Leitenberg, *Assessing.*

12. For a detailed discussion of state biological weapons programs, see www.nti.org

13. After living with HIV/AIDS for nearly thirty years, it may be time to stop using the term *pandemic* and to begin thinking of HIV as an infectious agent that is endemic in the human population.

Glossary

Aerosol: A system of particles suspended in a gas or fog, dispersible in the air.

Agroterrorism: Terrorism involving the alteration or destruction of crops or domesticated farm animals.

Amerithrax: FBI code name for the post–September 11 anthrax attacks.

Antibiotic: Substance produced by one microbe that kills or disables other microbes.

Antibody: A protein found in blood, produced in response to invasion of the body by a microbe or other foreign biological entity, capable of recognizing that entity and promoting its elimination.

Antiserum: Serum (q.v.) containing antibody.

Antiviral (drug): Chemical substance interfering with reproduction or function of a virus.

Antitoxin: Antiserum made against a bacterial toxin.

Aquifer: A geological formation containing water.

Bacillus: Rod-shaped bacterium.

Bacteriophage: A virus that infects bacteria.

Bacterium: The structurally simplest single-cell organism.

Biocrime: The use of biological weapons for criminal purposes.

Biopreparat: Umbrella organization overseeing all bioweapons research in the Soviet Union.

Biosensor: A federal program for the design and placement of detectors for the collection and identification of biological agents used in terrorism.

Biotechnology: The mechanical and engineering technology used in the manufacture of medicines and medical devises.

Bioterrorism: The use of biological weapons for political, social, or economic terrorism.

Buboes: Largely obsolete term for enlarged lymph nodes.

CDC: Centers for Disease Control and Prevention.

Chromosome: Structure in the nucleus of a cell containing the cell's DNA and genes.

Cipro: Ciprofloxacin, a wide-spectrum antibiotic useful in treating anthrax.

Contagious: Capable of being passed between and among people and/or animals (said of a disease or disease agent).

DNA fingerprint: DNA sequences used to establish identity (or non-identity).

DNA sequencing: Determination of the nucleotide sequence in DNA.

DNA vaccine: A vaccine in which microbial DNA is used rather than the microbe itself. The host produces microbial proteins from the DNA, which then induce an immune response in the host.

Edema: Local swelling resulting from failure to drain lymph fluid from tissue.

Epidemic: A sudden, rampant spread of an infectious disease among humans within a single country, or possibly adjoining countries.

Epizootic: Epidemic in animal populations.

Feedlot: Location where animals are maintained in close quarters and fed in preparation for slaughter.

Gene: Unit of DNA encoding information for production of a protein or RNA.

Genetic engineering: The planned alteration of a DNA sequence.

Genome: The entire set of DNA sequences defining an individual organism, used to direct that organism's reproduction and function.

Germination: The conversion of a plant seed or bacterial spore to a viable organism.

Hemorrhagic fever: Inflammatory state, usually virally induced, accompanied by excessive bleeding.

HIV: Human immunodeficiency virus.

Immunopathology: Disease caused by the immune system itself.

Incubation period: Period between infection and first appearance of clinical symptoms.

Infectious disease: Disease caused by infectious, pathogenic microbes.

Isolation: Confinement of an individual known to have a contagious disease.

MDR-TB: Moderately drug-resistant tuberculosis.

Microbe: A single-cell living organism.

Microbiology: The study of microbes.

Microorganism: A living organism that can only be seen with the aid of a microscope.

Molecular biology: The chemistry and biology of DNA.

Morbidity: Illness.

Mutation: An induced or natural alteration in the sequence of DNA.

Natural immunity: Immunity acquired as a result of natural exposure to a microbe.

Neurotoxin: A toxin altering function of a nerve cell.

Nucleotide: The basic chemical building block of DNA.

Pandemic: Spread of a disease through a large number of countries, or more than one continent.

Pathogen: A biological agent causing disease.

Pathogenic: Causing disease.

Prophylactic: Preventative.

Reservoir: In epidemiology, an animal species that can harbor a human pathogen without itself contracting a lethal form of the disease caused by that pathogen in humans.

Respirator: A protective mask to limit intake of harmful substances through the airways.

Quarantine: Confinement of an individual presumed to have a contagious disease.

Recombinant DNA: DNA artificially created by chemically combining DNA from two different sources.

Recombinant protein: Protein made from recombinant DNA.

SARS: Severe acute respiratory syndrome.

Septicemia: Large numbers of bacteria in the bloodstream.

Sepsis: Condition in which microbes (usually bacteria) replicate within the body with no control by the immune system.

SNS: Strategic National Stockpile.

Social distancing: Maximal spacing of people to reduce person-person transmission of disease.

Spore: Hard, dry, compacted form of a bacterial cell.

Surge capacity: Ability to increase any critical component of a coordinated response.

Synthetic biology: Creation of new life forms from existing forms, or from scratch.

Tamiflu: A commercially available anti-flu drug (trade name of oseltamivir phosphate).

Toxic shock: Physiological shock caused by release of bacterial toxins.

Toxin: In microbiology, a chemical released by a parasitic microbe that is harmful to its host.

Triage: Sorting, especially of patients according to their need for care.

Vaccination: Immunization with all or part of a pathogen, intended to build immunity before natural exposure to the pathogen.

Vaccine: A killed microbe, or some portion thereof, used to provoke protective immunity to that microbe.

Ventilator: Mechanical device to assist breathing.

Virulent: Highly infectious.

WMD: Weapons of mass destruction (biological, chemical, or nuclear).

XDR-TB: Extremely drug-resistant tuberculosis.

INDEX